International

REVIEW OF

Neurobiology

Volume 89

International REVIEW OF *Neurobiology*

Volume 89

SERIES EDITORS

RONALD J. BRADLEY
Department of Psychiatry, College of Medicine
The University of Tennessee Health Science Center
Memphis, Tennessee, USA

R. ADRON HARRIS
Waggoner Center for Alcohol and Drug Addiction Research
The University of Texas at Austin
Austin, Texas, USA

PETER JENNER
Division of Pharmacology and Therapeutics
GKT School of Biomedical Sciences
King's College, London, UK

EDITORIAL BOARD

Novel Approaches to Studying Basal Ganglia and Related Neuropsychiatric Disorders

EDITED BY

XIAO-HONG LU

Brain Research Institute
Semel Institute for Neuroscience and Human Behavior
Center for Neurobehavioral Genetics
Department of Psychiatry and Biobehavioral Sciences
David Geffen School of Medicine at UCLA
Gonda, Los Angeles,
California, USA

ELSEVIER

AMSTERDAM • BOSTON • HEIDELBERG • LONDON
NEW YORK • OXFORD • PARIS • SAN DIEGO
SAN FRANCISCO • SINGAPORE • SYDNEY • TOKYO
Academic Press is an imprint of Elsevier

Academic Press is an imprint of Elsevier
360 Park Avenue South, New York, NY 10010-1700
525 B Street, Suite 1900, San Diego, California 92101-4495, USA
32 Jamestown Road, London NW1 7BY, UK

This book is printed on acid-free paper.

For information on all Academic Press publications
visit our Web site at www.elsevierdirect.com

ISBN-13: 978-0-12-374894-2

PRINTED AND BOUND IN THE UNITED STATES OF AMERICA
09 10 11 12 8 7 6 5 4 3 2 1

CONTENTS

Behavioral Outcome Measures for the Assessment of Sensorimotor Function in Animal Models of Movement Disorders

Sheila M. Fleming

The Role of DNA Methylation in the Central Nervous System and Neuropsychiatric Disorders

Jian Feng and Guoping Fan

Heritability of Structural Brain Traits: An Endophenotype Approach to Deconstruct Schizophrenia

Nil Kaymaz and J. van Os

The Role of Striatal NMDA Receptors in Drug Addiction

Yao-Ying Ma, Carlos Cepeda, and Cai-Lian Cui

Deciphering Rett Syndrome With Mouse Genetics, Epigenomics, and Human Neurons

JIFANG TAO, HAO WU, AND YI EVE SUN

CONTRIBUTORS

Numbers in parentheses indicate the pages on which the authors' contributions begin.

Carlos Cepeda (131), Mental Retardation Research Center, Semel Institute for Neuroscience and Human Behavior, University of California Los Angeles, Los Angeles, California, USA

Cai-Lian Cui (131), Neuroscience Research Institute and Department of Neurobiology, Peking University Health Science Center, People's Republic of China

Guoping Fan (67), Department of Human Genetics, David Geffen School of Medicine, University of California, Los Angeles, California 90095, USA

Jian Feng (67), Department of Human Genetics, David Geffen School of Medicine, University of California, Los Angeles, California 90095, USA

Sheila M. Fleming (57), Departments of Psychology and Neurology, University of Cincinnati, Cincinnati, Ohio 45221, USA

Nil Kaymaz (85), Mediant GGZ/Mental Health Care, Postbus 775, 7500 AT Enschede, The Netherlands; and Department of Psychiatry and Psychology, University of Maastricht, P.O. Box 616 (DRT10) Maastricht, The Netherlands

Mary Kay Lobo (1), Department of Neuroscience, Mount Sinai School of Medicine, New York, NY, USA

Xiao-Hong Lu (37), Department of Psychiatry and Biobehavioral Sciences, Center for Neurobehavioral Genetics, Semel Institute for Neuroscience and Human Behavior, Brain Research Institute, David Geffen School of Medicine at UCLA, Los Angeles, California 90095, USA

Yao-Ying Ma (131), Stefan and Shirley Hatos Center for Neuropharmacology, University of California Los Angeles, Los Angeles, California, USA

J. van Os (85), Division of Psychological Medicine, Institute of Psychiatry, London, United Kingdom; and Department of Psychiatry and Psychology, University of Maastricht, P.O. Box 616 (DRT10) Maastricht, The Netherlands

Yi Eve Sun (147), Department of Molecular and Medical Pharmacology and Psychiatry and Behavioral Sciences, University of California, Los Angeles, USA

Jifang Tao (147), Department of Molecular and Medical Pharmacology and Psychiatry and Behavioral Sciences, University of California, Los Angeles, USA

Hao Wu (147), Department of Molecular and Medical Pharmacology and Psychiatry and Behavioral Sciences, University of California, Los Angeles, USA

PREFACE

NOVEL APPROACHES TO STUDYING BASAL GANGLIA AND RELATED NEUROPSYCHIATRIC DISORDERS

Basal ganglia are a large collection of subcortical nuclei interconnected with the cerebral cortex, thalamus, and brainstem. Physiological function of basal ganglia not only related to movement control but also has been implicated in cognitive function, emotion, and learning. A lot of human basal ganglia-related disorders, such as Huntington's disease (HD), Parkinson's disease (PD), drug addiction, Tourette's syndrome, obsessive–compulsive disorder, Rett syndrome, and schizophrenia, should be viewed as dysfunction of specific cortico-subcortical loops. These circuits comprise reentrant pathways that originate in various neocortical domains and pass through the basal ganglia before returning, by way of the thalamus, to select portions of the frontal lobe. The major revisions of the views about anatomy and function (dysfunction) of basal ganglia in the last decade demand new perspectives and methodology to dissect the functions of the circuitries and to understand the role of these structures in various neuropsychiatric disorders. Recently, many novel approaches have been used to dissect biology and pathophysiology of basal ganglia. These approaches include FACS-array molecular profiling, bacterial artificial chromosome (BAC) transgenic mouse models, DNA methylation, epigenomics, neuroplasticity, novel behavioral analysis, and human genetics. The current volume will review some advances in these rapidly developing areas and try to provide new insights into the study of the pathophysiology of basal ganglia-related disorders and enhance our understanding of the function of the basal ganglia in movement control, action selection, and reinforcement learning.

XIAO-HONG LU

MOLECULAR PROFILING OF STRIATONIGRAL AND STRIATOPALLIDAL MEDIUM SPINY NEURONS: PAST, PRESENT, AND FUTURE

Mary Kay Lobo

Department of Neuroscience, Mount Sinai School of Medicine, New York, NY, USA

Defining distinct molecular properties of the two striatal medium spiny neurons (MSNs) has been a challenging task for basal ganglia (BG) neuroscientists. Identifying differential molecular components in each MSN subtype is crucial for BG researchers to understand functional properties of these two neurons. The two MSN populations are morphologically identical except in their projections through the direct verses indirect BG pathways and they are heterogeneously dispersed throughout the dorsal striatum (dStr) and nucleus accumbens (NAc). These characteristics have made it difficult for researchers to distinguish and isolate these two neuronal populations thereby hindering progress toward a more comprehensive understanding of their differential molecular properties. Researchers began to investigate molecular differences in the striatonigral and striatopallidal neurons using *in situ* hybridization (ISH) techniques and single cell reverse transcription-polymerase chain reaction (scRT-PCR). Currently the field is utilizing more advanced techniques for large-scale gene expression studies including fluorescence activated cell sorting (FACS) of MSNs, from which RNA is purified, from fluorescent reporter transgenic mice or use of transgenic mice in which ribosomes from each MSN are tagged and can be immunoprecipitated followed by RNA isolation, a technique termed translating ribosomal affinity purification (TRAP). Additionally,

INTERNATIONAL REVIEW OF NEUROBIOLOGY, VOL. 89
DOI: 10.1016/S0074-7742(09)89001-6

the availability of fluorescent reporter mice for each MSN subtype is allowing, scientists to perform more accurate histology studies evaluating differential protein expression and signaling changes in each cell subtype. Finally, researchers are able to evaluate the role of specific genes *in vivo* by utilizing cell type-specific mouse models including Cre driver lines that can be crossed with conditional overexpression or knockout systems. This is a very exciting time in the BG field because researchers are well equipped with the most progressive tools to comprehensively evaluate molecular components in the two MSNs and their consequence on BG functional output in the normal, diseased, and developing brain.

I. Introduction

The current model of the basal ganglia (BG) involves two distinct pathways (the direct and indirect pathway) and they are thought to play antagonistic but balancing roles on the BG output and behavior (Albin *et al.*, 1989; DeLong and Wichmann, 2007; Graybiel, 2000). The two pathways can be resolved at a cellular level in the main projection neurons of the striatum. The projection neurons, which comprise 90–95% of all neurons in the striatum, are these medium spiny neurons (MSNs) (Fig. 1A), which are divided into two morphologically identical and heterogeneously distributed cell types (Fig. 1B) (Gerfen, 1992; Nakano *et al.*, 2000). The direct pathway is composed of the striatonigral MSNs which project to the globus pallidus interna (GPi) and substantia nigra (SN) while the indirect pathway is composed of striatopallidal MSNs which project to the globus pallidus externa (GPe) and indirectly influence the GPi and SN via the subthalamic nucleus (STN) (Fig. 1C) (Gerfen, 1992). This model was primarily associated and most well studied with movement output and movement disorders in the BG, which include Huntington's Disease and Parksinon's Disease. The striatonigral MSNs in dStr promote movement by causing disinhibition of excitatory thalamic neurons which project to motor cortex, in contrast the striatopallidal MSNs in dStr causing inhibition of movement by inhibiting excitatory thalamic neurons (Albin *et al.*, 1989; DeLong and Wichmann, 2007; Graybiel, 2000). Some studies involving genetic ablation of either striatonigral or striatopallidal MSNs support this model in motor behaviors (Drago *et al.*, 1998; Sano *et al.*, 2003; Durieux *et al.*, 2009). Despite this models known role in motor output it is important to recognize the potential role for this model in other behaviors, such as reward and emotion, habit, learning and memory, and cognition (Belin *et al.*, 2008; Graybiel, 2005; Packard and Knowlton, 2002; Yin and Knowlton,

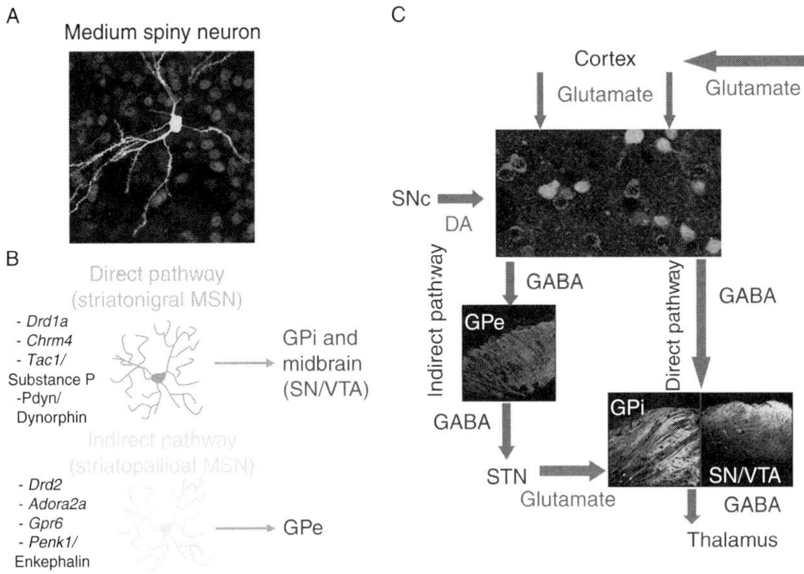

FIG. 1. Medium spiny neurons and the direct and indirect pathways. (A) Image of a medium spiny neuron (MSN) in green with red background nuclear staining. (B) Illustration of the two MSN subtypes. Striatonigral MSN, which constitute the direct pathway, send projections to the globus pallidus interna (GPi), substantia nigra (SN), and ventral tegmental area (VTA). Whereas, striatopallidal MSNs constitute the indirect pathway and send projections to globus pallidus externa (GPe). Striatonigral MSNs are enriched in *Drd1a*, *Chrm4*, *Tac1* (gene for substance P), and *Pdyn* (gene for dynorphin) and striatopallidal MSNs are enriched in *Drd2*, *Adora2a*, *Gpr6*, and *Penk1* (gene for enkephalin). (C) Diagram of the BG circuit. In the direct pathway striatonigral MSNs send projections to GPi and SN, which project to thalamus, and activation of this pathway can increase cortical drive via disinhibition of thalamus. Striatopallidal MSNs in the indirect pathway project to GPe, which feeds into the direct pathway via the subthalamic nucleus (STN), and activation of this pathway can decrease cortical drive via inhibition of the thalamus. Neurotransmitters, GABA and glutamate, are noted in the diagram. Also noted are the dopamine (DA) inputs from substantia nigra pars compacta (SNc), which have modulatory influence on the two pathways. (See Color Insert.)

2006) since striatum receives input from many cortical areas resulting in multiple cortico-basal ganglia-cortico loops (DeLong and Wichmann, 2007; Graybiel, 2000; Herrero *et al.*, 2002; Nakano, 2000). An imbalance of the two BG pathways in these behaviors has been implicated in neuropsychiatric disorders such as drug addiction, obsessive compulsive disorders (OCD), Tourette's syndrome (TS), and attention deficit hyperactive disorder (ADHD) (Chao and Nestler, 2004; Graybiel and Rauch, 2000; Saka and Graybiel, 2003). Despite these insights there is still a need for a more comprehensive understanding of the functional role of these two pathways and molecular mechanisms in the two MSN subtypes in BG behaviors and associated diseases.

Although this chapter is focused on the two MSNs in the direct and indirect pathways it is important to note further subdivisions of the striatum. The striatum is further divided into dorsal striatum (dStr) and ventral striatum/nucleus accumbens (NAc). The dStr which receives inputs from associative cortex and sensorimotor cortex (Herrero *et al.*, 2002; Nakano, 2000) is most known for its role on motor function (Albin *et al.*, 1989; DeLong and Wichmann, 2007; Graybiel, 2000) whereas the NAc receives input from orbitofrontal cortex and other limbic cortical areas (Herrero *et al.*, 2002; Nakano, 2000) and is primarily known for its role in emotional and reward behaviors (Belin *et al.*, 2008; Cardinal *et al.*, 2002; Hyman *et al.*, 2006; Pecina *et al.*, 2006). Research has been focused on evaluating differences in the two MSNs in both the dStr and NAc. Analyzing differential molecular mechanisms in the two neuronal subtypes in both regions has implications for a variety of disorders including movement disorders and drug addictions.

Additionally, the dStr can further be subdivided into the patch compartment (or striosomes) and the matrix compartment. The functional role of these compartments is still poorly understood but the patch compartment has been shown to receive input from frontal and limbic cortical areas and the matrix from somatosensory and motor cortex (Gerfen, 1984). Each compartment projects to different areas of the SN, the patch neurons project to dopamine (DA) neurons in the substantia nigra pars compacta (SNc) and the matrix neurons project to the GABAergic neurons in the substantia nigra reticulata (SNr) (*ibid.*). Furthermore, each compartment displays a different time course of development (Fishell and van der Kooy, 1987, 1989, 1991; van der Kooy and Fishell, 1987) and there is a large area of research devoted to understanding critical molecular components involved in development of patch versus matrix compartments (Anderson *et al.*, 1997; Garel *et al.*, 1999; Mason *et al.*, 2005; Toresson and Campbell, 2001; Yun *et al.*, 2003). In fact, the majority of research on molecular development of the striatum is in the context of these two compartments but little is known about the differential development of the two striatal MSNs. Recent evidence suggests that striatonigral MSNs in the matrix compartment may have differential developmental processes than the striatonigral MSNs in the patch compartment (Lobo *et al.*, 2008). It is plausible that different developmental processes also occur in the striatopallidal patch and matrix MSNs as well. Understanding molecular properties of the two MSNs is important for understanding how these two neurons develop during pre- and postnatal periods and may help better understand developmental BG disorders such as TD and ADHD.

When discussing the two BG pathways it is imperative to address the role of DA in the two (Le Moine *et al.*, 1990) MSNs and BG output. DA is known for its important role in motor and reward behaviors in the BG (Berridge and Robinson, 1998; DeLong and Wichmann, 2007; Schultz, 2002) and dysfunction of DA signaling is implicated in motor and psychiatric diseases (DeLong and Wichmann, 2007; Hyman *et al.*, 2006). dStr and NAc receive DA inputs from the SNc and ventral tegmental area (VTA), respectively. DA Receptor 1a (*Drd1a*) and DA receptor 2 (*Drd2*) are the most enriched DA receptors in the striatum (Table I) and it is well established in the BG field that striatonigral MSNs are enriched in *Drd1a* receptors and striatopallidal MSNs are enriched in *Drd2* receptors (Gerfen *et al.*, 1990; Gerfen, 1992; Le Moine *et al.*, 1990, 1991; Surmeier *et al.*, 1996). DA signaling in striatum is thought to facilitate excitatory glutamatergic input from cortical areas onto striatonigral MSNs. This occurs through *Drd1a* receptor signaling, which activates adenylyl cyclase via G_{olf} leading to increased cytosolic cAMP levels and increased protein kinase A (PKA) activity. PKA activity via *Drd1a* signaling has been shown to play a role in increased membrane surface expression of AMPA and NMDA receptors, inactivation of potassium channels, and opening of L-type Ca^{2+} channels all which support a role for DA facilitation of cortical glutamatergic signaling onto striatonigral MSNs in the direct pathway(Missale *et al.*, 1998; Surmeier *et al.*, 2007). In contrast DA signaling via *Drd2*, in striatopallidal MSNs, through $G_{i/o}$ inhibits adenylyl cyclase but results in increased intracellular Ca^{2+} stores and increased protein kinase C (PKC) activity. This signaling cascade has been shown to negatively regulate Cav1.3 Ca^{2+} channels, decrease opening of voltage dependent sodium channels, and increase K^+ channel opening supporting DA attenuation of excitatory cortical input onto striatopallidal MSNs in the indirect pathway (*ibid.*). These findings substantiate a role for DA to facilitate the direct pathway while putting a brake on the indirect pathway thereby resulting in increased cortical drive via increased disinhibition of excitatory thalamocortical neurons. Understanding the differential DA signaling properties in the two MSNs and BG output has been essential for understanding differential functional roles of these two neurons on the BG in the normal and diseased brain.

Despite the critical role of DA signaling in the two MSN subtypes, the differential expression of G-protein coupled receptors in the two neurons is not limited to DA receptors. Striatonigral MSNs are also enriched in muscarinic receptor M4 (*Chrm4*) and striatopallidal receptors are enriched in Adenosine 2A receptor (*Adora2A*) and G-protein coupled receptor 6 (*Gpr6*) (Table I) (Bernard *et al.*, 1992; Ince *et al.*, 1997; Lobo *et al.*, 2007; Santiago and Potter, 2001; Schiffmann and Vanderhaeghen, 1993; Schiffmann *et al.*, 1991a). Finally, these two neurons differentially express neuropeptides, substance P (*Tac1*—the gene for substance P) and dynorphin (*Pdyn*—the gene for dynorphin) in the striatonigral MSNs and enkephalin (*Penk1*—the gene for enkephalin) in striatopallidal MSNs (Table I) (Gerfen and Young, 1988).

Researchers first began to profile these cells using *in situ* hybridization (ISH) combined with retrograde tracing studies, double ISH, or double ISH/immunohistochemistry (Fig. 2A,B) to determine gene expression of *Drd1a*, *Chrm4*, *Tac1*, and *Pdyn* in striatonigral MSNs and *Drd2*, *Adora2A*, and *Penk1* in striatopallidal MSNs (Table I). This was further validated using single cell (scRT-PCR) (Fig. 2C).

TABLE I

G-protein coupled receptors and neuropeptides enriched in each MSN subtype. List of GPCRs and neuropeptide genes enriched in striatonigral and striatopallidal MSNs

	Striatonigral MSNs (Direct pathway)		
	Gene name	Protein name	Original studies describing enrichment in either MSN
Gpcrs	Dopamine receptor 1a (*Drd1a*)	*	C. R. Gerfen *et al.*, 1990; C. Le Moine *et al.*, 1991
	Acetylcholine muscarinic receptor 4 (*Chrm4*)	*	V. Bernard *et al.*, 1992; E. Ince *et al.*, 1997
Neuropeptides	Tachykinin 1 (*Tac1*)	Substance P	C. R. Gerfen and W. S. Young, 3rd, 1988
	Prodynorphin (*Pdyn*)	Dynorphin	C. R. Gerfen and W. S. Young, 3rd, 1988

	Striatopallidal MSNs (Indirect pathway)		
	Gene name	Protein name	Original studies describing enrichment in either MSN
Gpcrs	Dopamine receptor 2 (*Drd2*)	*	C. R. Gerfen *et al.*, 1990; C. Le Moine *et al.*, 1990
	Adenosine receptor 2A (*Adora2a*)	*	S. N. Schiffmann *et al.*, 1991a
			S. N. Schiffmann and J. J. Vanderhaeghen, 1993
	G protein coupled receptor 6 (*Gpr6*)	*	M. K. Lobo *et al.*, 2007; www.gensat.org
Neuropeptide	Preproenkephalin (*Penk1*)	Enkephalin	C. R. Gerfen and W. S. Young, 3rd, 1988

*Protein name is the same as the gene name.

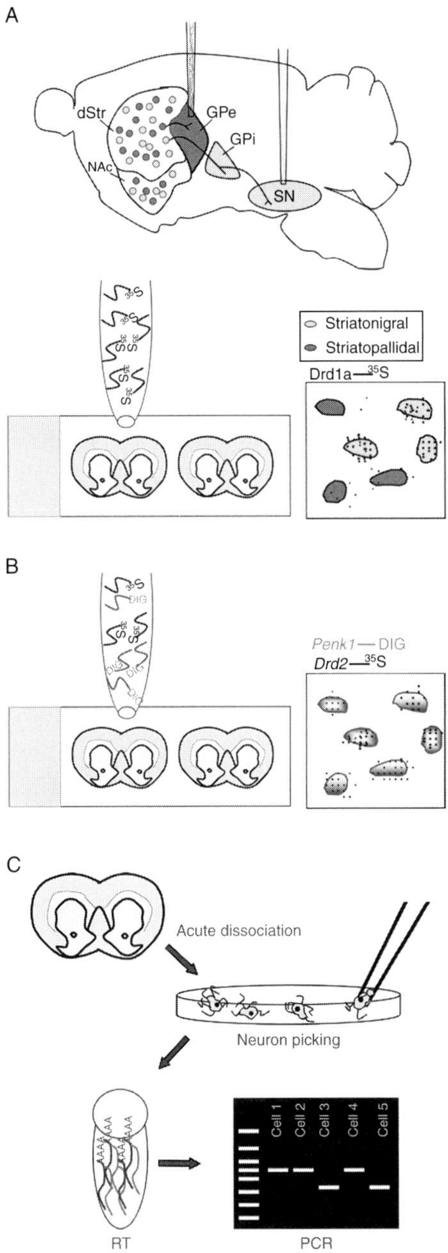

FIG. 2. *In situ* hybridization (ISH) and scRT-PCR techniques to evaluate gene expression in each MSN population. (A) Retrograde labeling of either GPe or SN combined with radioactive ^{35}S ISH can be used to evaluate gene expression in striatonigral (yellow retrograde labeling from SN) or

At the present time, we are currently able to profile large gene sets in the two MSNs with the help of BAC transgenic mice expressing cytosolic EGFP or ribosomal tagged EGFP using *Drd1a*, *Chrm4*, and *Drd2* BACs and by utilizing techniques (FACS and bac-TRAP followed by RNA purification) to isolate RNA in each MSN subtype (Heiman *et al.*, 2008; Lobo *et al.*, 2006). Additionally the *Drd1a*-, *Chrm4*-, and *Drd2*-EGFP BAC transgenic mice have recently been used to analyze neurophysiological disparities and differential signaling properties using immunofluorescence staining in the two MSNs (Ade *et al.*, 2008; Bertran-Gonzalez *et al.*, 2008, 2009; Borgkvist *et al.*, 2008; Brami-Cherrier *et al.*, 2005; Cepeda *et al.*, 2008; Day *et al.*, 2006; Ding *et al.*, 2008; Gertler *et al.*, 2008; Kim *et al.*, 2009; Kreitzer and Malenka, 2007; Lee *et al.*, 2006; Santini *et al.*, 2007; Shen *et al.*, 2007, 2008; Taverna *et al.*, 2008; Valjent *et al.*, 2000; Wang *et al.*, 2006b). This review attempts to address past and current techniques to profile the molecular machinery specific to the two MSNs and the implications these findings have had on function and development of the two MSNs in the BG circuit. Finally we are poised at a very exciting time point for molecular profiling and the understanding of differential molecular mechanisms in the two MSNs. We have the appropriate tools and knowledge to truly dissect out molecular disparities in these two neurons. This review will also address other current and future techniques and experiments, such as chromatin immunoprecipitation (ChIP) and use of mice expressing Cre-recombinase in each MSN population combined with conditional systems. Researchers are in a position to utilize these techniques to gain a more comprehensive knowledge of the molecular properties in the two MSNs and their functional and behavioral outcomes in the normal, diseased, and developing BG circuit.

II. Early Detection of Neuropeptides and GPCRs in Striatonigral and Striatopallidal MSNs

Unraveling molecular and neurochemical differences in the two MSN subtypes has been an ongoing process for the past 30 years. The first indication that there was a neurochemical difference in striatal neurons that project to the GPi

striatopallidal (green retrograde labeling from GPe) MSNs. Cartoon example of Drd1a ^{35}S ISH (black dots) colocalizing with striatonigral MSNs (yellow cells). (B) Double ISH with ^{35}S and DIG probes can be used to distinguish two different genes simultaneously. Cartoon example of Penk1 (DIG labeled purple cell) colocalizing with Drd2 (^{35}S black dots). (C) Cartoon image of scRT-PCR technique, which can be used to profile multiple genes in one neuron. Striatal tissue is acutely dissociated, single neurons are aspirated into fine glass tips then placed in a microcentrifuge tube, and mRNA in the cell is converted into cDNA using a reverse transcription (RT) reaction. cDNA can be used for PCR then run on a DNA electrophoresis gel to visualize the PCR product.

and SN versus those that project to the GPe came about when researchers investigated the origination of the substance P and dynorphin protein expressing terminals in the SN and enkephalin protein terminals in the GPe. Techniques such as striatal lesions or knife cutting of striatal projections to the SN or GPe in rodents combined with radioimmunoassays and immunohistochemistry showed that the striatonigral MSNs contain substance P and dynorphin (Brownstein *et al.*, 1977; Hong *et al.*, 1977b; Kanazawa *et al.*, 1977; McLean *et al.*, 1985; Mroz *et al.*, 1977) and striatopallidal MSNs contain enkephalin (Cuello and Paxinos, 1978; Del Fiacco *et al.*, 1982; Hong *et al.*, 1977a). The substance P data were further supported by experiments using *in vivo* radiolabeled synthesis of substance P in rats to confirm that striatonigral MSNs synthesize substance P and it is transported to the SN via striatonigral axonal projections (Krause *et al.*, 1984). These data are important evidence, which originally suggested differential molecular components in the striatonigral verses striatopallidal MSNs. Gerfen and colleagues pursued this differential neuropeptide expression by using a combination of retrograde axonal tracing and ISH to demonstrate that the majority of neurons expressing enkephalin (*Penk1* mRNA) project to the GPe whereas the majority of substance P and dynorphin (*Tac1* and *Pdyn* mRNA) containing MSNs project to the SN (Gerfen and Young, 1988). This work set the stage for gene expression analysis in the two MSN subtypes. Throughout the next decade researchers would continue to use ISH techniques (Fig. 2A,B) to quantify gene expression of other molecules in the striatonigral and striatopallidal MSNs. Additionally, expression of *Tac1* (or *Pdyn*) and *Penk1* in each MSN subtype would be used to validate each cell type population in ISH experiments and later on using scRT-PCR experiments.

The next step to analyze molecular disparities in the two MSNs was focused on DA receptors. Research supported expression of DA receptors in striatum (Forn *et al.*, 1974) and that both classes of DA receptors, D1 class (*Drd1a* and *Drd1b*) and D2 class (*Drd2*, *Drd3*, and *Drd4*) (Lachowicz and Sibley, 1997) are expressed in striatum (Creese *et al.*, 1983; Kebabian and Calne, 1979). To address the composition of D1 class and D2 class receptors in the two MSN subtypes researchers performed ISH for *Drd1a* or *Drd2* receptor mRNA combined with retrograde axonal tracing (Gerfen *et al.*, 1990) or double ISH for Drd1a or Drd2 receptor mRNA with simultaneous detection of *Tac1* or *Penk1* mRNA (Le Moine *et al.*, 1990, 1991). These data revealed an enrichment of *Drd1a* in *Tac1* containing striatonigral MSNs and enrichment of *Drd2* in *Penk1* expressing striatopallidal MSNs. Additionally, 6-OHDA lesions of DA neurons in the SN resulted in an increase of striatopallidal *Drd2* and *Penk1* expression but a decrease in *Drd1a* and *Tac1* expression implicating a role for DA in regulating mRNA expression in each MSN subtype (Gerfen *et al.*, 1990). Discovering differential DA receptor composition and signaling in the two MSNs was an important milestone for future understanding of signaling differences and BG circuit output in the two cell subtypes.

Despite the importance of DA signaling in the two neurons it is important to address other GPCR signaling mechanisms since differential GPCR expression is not limited to DA receptors in the two MSNs. Evidence for adenosine G-protein signaling, expression of adenosine receptors types A1 and A2 in striatum (Wojcik and Neff, 1983a,b,c) and the data showing *Adora2a* is restricted to striatum (Schiffmann *et al.*, 1990, 1991b) prompted researchers to investigate differential expression of adenosine receptors in the two MSNs. ISH studies using adjacent sections hybridized for *Adora2a*, *Penk1*, *Tac1*, *Drd1a*, or *Drd2* and combining ISH with retrograde tracing from the SN revealed that *Adora2A* is enriched in striatopallidal but not striatonigral MSNs (Schiffmann and Vanderhaeghen, 1993; Schiffmann *et al.*, 1991a). These data laid a platform for analysis of adenosine signaling in striatum, specifically through *Adora2a* receptors in striatopallidal MSNs. *Adora2a* receptors are G_s coupled and promote increases in cAMP in striatopallidal MSNs which is opposite to *Drd2* receptor function in these neurons (Svenningsson *et al.*, 1999), therefore *Adora2A* receptors promote this braking indirect pathway in contrast to *Drd2* receptors which are thought to play an inhibitory role in this pathway. The relevance of *Adora2a* receptors in the striatopallidal MSNs has had much attention in the past decade with focus on *Adora2a* antagonists for Parkinson's Disease treatment (Richardson *et al.*, 1997; Schwarzschild *et al.*, 2006) where loss of DA is thought to lead to an overactive indirect pathway (Albin *et al.*, 1989; DeLong and Wichmann, 2007).

Acetylcholine receptors are another class of GPCRs that are of interest in striatal function. Muscarinic acetylcholine receptors play important roles in striatal neuron function and BG output (Calabresi *et al.*, 2000). Three muscarinic receptors (*Chrm1*, *Chrm2*, and *Chrm4*) were found to be enriched in striatum (Buckley *et al.*, 1988; Vilaro *et al.*, 1991; Weiner *et al.*, 1990) and *Chrm4* was shown to be enriched in striatonigral MSNs using ISH on adjacent sections or double ISH (Bernard *et al.*, 1992) or by SN retrograde tracing and immunohistochemistry (Ince *et al.*, 1997). This colocalization was further verified using retrograde tracing of SN and GPe combined with visualization of Biotinylated *Chrm4*-toxin complexes which selectively bind *Chrm4* receptors (Santiago and Potter, 2001).

The majority of the techniques to determine differential gene expression in the two MSNs involved ISH and/or retrograde tracing. However, in the late 1990s researchers began profiling these neurons using scRT-PCR a technique originally used by Eberwine *et al.* (1992) in which acutely dissociated single neurons are aspirated into fine glass pipette tips and RNA from these neurons subsequently undergo RT (Fig. 2C). This approach was used to verify whether substance P (*Tac1*) or enkephalin (*Penk1*) was localized to the same neurons (Surmeier *et al.*, 1996; Wang *et al.*, 2006a) and more comprehensively evaluate *Drd1a* and *Drd2* receptor localization to substance P and enkephalin expressing neurons (Surmeier *et al.*, 1996). This approach revealed that a small subset of neurons coexpress both *Tac1* and *Penk1* and/or *Drd1a* and *Drd2* in rodents at ages

3–4 weeks (Surmeier *et al.*, 1996; Wang *et al.*, 2006a) with the number declining in adult rats at age 4 months (Wang *et al.*, 2006a). Additionally, it was shown that other D1 class (*Drd1b*) and D2 class receptors (*Drd3* and *Drd4*) are expressed in both cell types, albeit in lower levels, implicating a role for D1 class and D2 class receptor signaling in both MSNs (Surmeier *et al.*, 1996). Furthermore, scRT-PCR was used to validate expression of muscarinic receptors and the data revealed the *Chrm1* was expressed in all MSNs as previously shown; *Chrm4* was expressed in both but more enriched in substance P containing neurons consistent with previous data; and *Chrm2*, *Chrm3*, and *Chrm5* were infrequently expressed in MSNs (Yan *et al.*, 2001). Finally this same group used this technique to further profile potassium channels, calcium channels, and AMPA receptors in the two MSNs (Foehring *et al.*, 2000; Mermelstein *et al.*, 1998, 1999; Stefani *et al.*, 1998). scRT-PCR is a powerful technique allowing researchers to detect the presence or absence of various genes in one cell and it is very useful for neurophysiologist to be able to correlate gene expression with functional properties in a single neuron. However, it can be extremely time consuming and is a difficult technique to use for large-scale microarray profiling. Later on in this chapter we will discuss current methods to analyze large-scale genetic differences in these two MSNs.

Once the set of known genes (GPCRs and neuropeptides) was identified researchers utilized these genes to evaluate changes in the two MSNs in neurological and neuropsychiatric diseases. In Huntington's Disease it is thought that striatopallidal MSNs are more vulnerable and preferentially degenerate in the disease (Mitchell *et al.*, 1999), which is consistent with the BG circuit model of an overactive direct pathway promoting increased motor behavior in this disorder (Albin *et al.*, 1989; Graybiel, 2000). Data showing a decrease in enkephalin expression by either immunohistochemistry or ISH in HD patients or HD mouse models support this assumption of a vulnerable striatopallidal indirect pathway (Menalled *et al.*, 2000; Sun *et al.*, 2002). Furthermore, researchers have evaluated the two MSNs in the NAc in reward behaviors. In particular, the role of ΔFosB in striatonigral and striatopallidal MSNs after chronic drug abuse or stress has been assessed. ΔFosB is a persistent, stable protein isoform of the FosB gene that accumulates in the NAc and dStr after repeated administration of drugs of abuse, chronic stress, and exposure to natural rewards (McClung *et al.*, 2004; Nestler, 2008; Nestler *et al.*, 2001; Perrotti *et al.*, 2004; Wallace *et al.*, 2008; Werme *et al.*, 2002). Evaluation of ΔFosB protein after treatment with D1 or D2 class agonists or antagonists suggests ΔFosB induction is mediated through *Drd1a* signaling in striatonigral MSNs (Moratalla *et al.*, 1996; Nye *et al.*, 1995). Furthermore, ISH/immunohistochemistry analysis reveals an increase in ΔFosB in both striatonigral and striatopallidal MSNs after chronic stress (Perrotti *et al.*, 2004) but only in striatonigral MSNs after chronic wheel running (a natural reward) (Werme *et al.*, 2002).

We have discussed techniques (double ISH, ISH/retrograde labeling, ISH/Immunohistochemistry, and scRT-PCR) researchers used to initially profile genetic differences in these two neurons. Those techniques have been incredibly useful to determine differential gene expression and protein expression in these two neurons and have led to important insights into how these neurons function differently in the BG circuit. However, these experiments have not been useful for large-scale gene expression studies. Furthermore, the labeling of the two cell types via retrograde labeling, ISH, and antibody staining has not yielded the best resolution so subtle differences between these two neurons might be missed. The remainder of this chapter will focus on current methods for large-scale genetic profiling of the two MSNs, current methods to reliably evaluate differential molecular mechanisms in each MSNs, and future techniques to comprehensively analyze genetic and functional differences between the two MSNs in the BG circuit.

III. Current Methods for Molecular Profiling in the Two MSN Subtypes

The BG field was in a dire need of a method to reliably differentiate the two MSNs. This problem has been solved in the last few years with the advent of BAC transgenic mice (Gong *et al.*, 2002; Heintz, 2001; Yang *et al.*, 1997), which led to high throughput genetic labeling of distinct neuronal populations using fluorescent proteins, gene expression nervous system atlas (GENSAT) (Gong *et al.*, 2003; http://www.gensat.org). Fortunately, for BG scientists, most of the known GPCRs and neuropeptide genes were candidates for generating BAC transgenic mice expressing EGFP using BACs for each gene. In particular researchers have begun to exploit BAC transgenic mice expressing EGFP using *Drd1a*, *Chrm4*, and *Drd2* BACS. These mice express EGFP in either striatonigral MSNs and their axonal projections to the GPi and SN (*Drd1a*-EGFP and *Chrm4*-EGFP BAC mice) or striatopallidal MSNs and their projections to the GPe (*Drd2*-EGFP BAC mice) (Fig. 3) (Bertran-Gonzalez *et al.*, 2008; Gong *et al.*, 2003; Lobo *et al.*, 2006; Matamales *et al.*, 2009; http://www.gensat.org). More recently another group has generated a *Drd1a*-tdTomato (a red fluorescent protein that is a derivative of GFP, Shaner *et al.*, 2004) BAC transgenic mouse (Shuen *et al.*, 2008) which can be crossed with the *Drd2*-EGFP mouse line to simultaneously visualize the two MSNs in one mouse. These fluorescent mice have been useful in validating expression of most of the genes known to be differentiated in the two MSNs (Gong *et al.*, 2003; Shuen *et al.*, 2008) and they are valuable tools for more comprehensive studies evaluating molecular and cellular differences in the two MSNs.

We utilized these mice, specifically the *Drd1a*-, *Chrm4*-, and *Drd2*-EGFP mice from GENSAT, to perform high throughput gene expression in the two MSN subtypes. We purified live EGFP neurons from juvenile and adult ages by

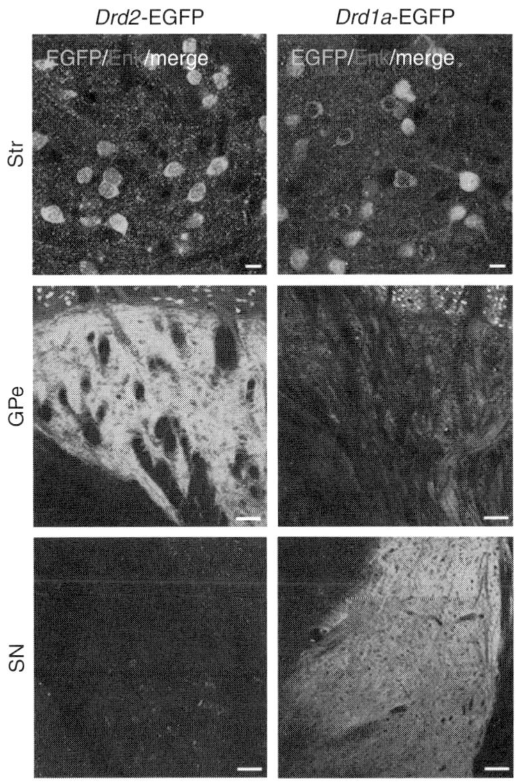

FIG. 3. EGFP expression in striatal MSNs and projections in *Drd2*-EGFP and *Drd1a*-EGFP BAC transgenic mice. *Drd2*-EGFP striatopallidal neurons (green, top left panel) colocalize with enkephalin positive neurons (red, top left panel) but *Drd1a*-EGFP striatonigral neurons (green, top right panel) do not coexpress enkephalin (red, top right panel). EGFP projections confirm that *Drd2*-EGFP striatopallidal MSNs display projection terminals to the GPe but not SN, whereas *Drd1a*-EGFP striatonigral MSNs display projection terminals to the SN but not GPe. Scale bars, top panels: 10 μm, bottom and middle panels: 100 μm. (See Color Insert.)

performing fluorescence activated cell sorting (FACS), then subsequently isolated RNA from the neurons for microarray profiling, which we have termed FACS-array (Fig. 4A) (Lobo *et al.*, 2006). Cell type-specific gene expression studies have become a huge focus of neuroscientists in recent years and researchers are using multiple techniques to dissect out neuronal populations in the very complex brain (Coppola and Geschwind, 2006; Nelson *et al.*, 2006). These techniques include a similar technique combining FACS with microarrays to isolate corticospinal motor neurons at embryonic and postnatal ages for gene expression profiling (Arlotta *et al.*, 2005), hand picking of single olfactory progenitors and adult

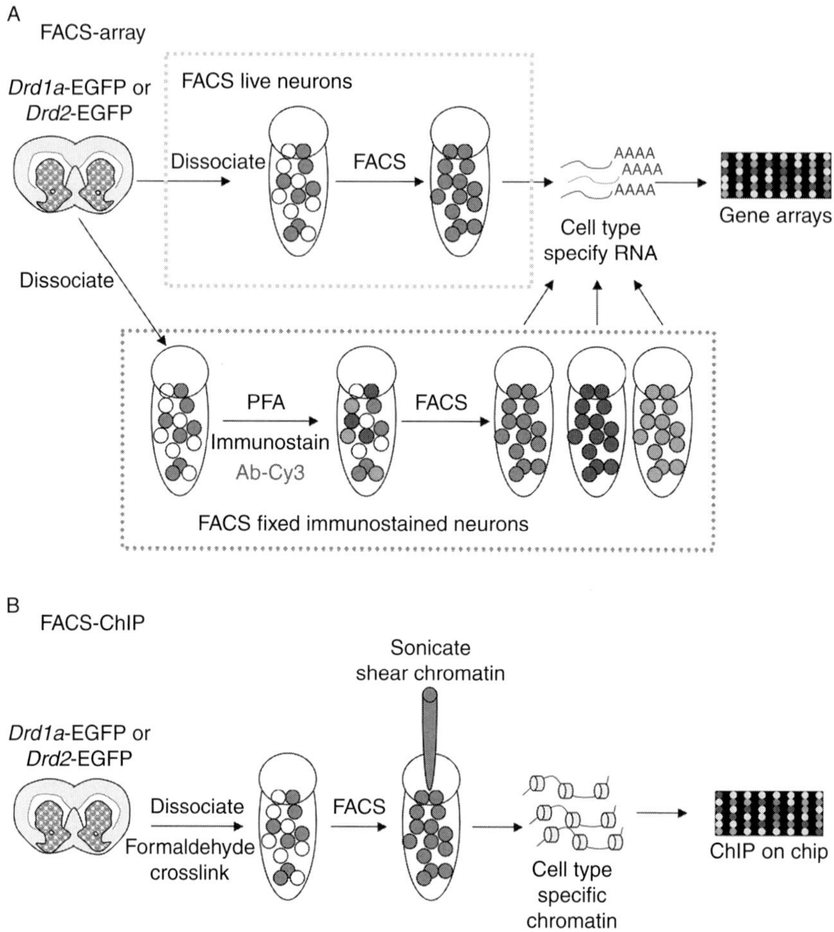

FIG. 4. Fluorescence activated cell sorting (FACS) to evaluate differential genetic identities in the two MSNs. (A) Diagram of FACS-array technique to profile gene expression changes in the two MSNs using transgenic mice that express EGFP in either MSN subtype. Immunostaining of EGFP neurons can be used to further isolate subpopulations within the two MSNs followed by gene expression changes. (B) Neuroscientists can now attempt to isolate chromatin from FACS purified MSNs to evaluate large-scale chromatin modifications in the two MSNs using ChIP or ChIP on chip. (See Color Insert.)

neurons for single cell microarray profiling (Tietjen *et al.*, 2003, 2005), single cell microarray of laser-capture microdissected CA1 hippocampal neurons (Kamme *et al.*, 2003, 2004), hand picking 12 populations of cortical neurons from adult ages for gene expression profiling (Sugino *et al.*, 2006), and more recently use of a novel technique translating ribosomal affinity purification (TRAP, this technique will be addressed later on in this section) to profile cell subtype genetic differences in 28 neuronal populations including striatonigral and striatopallidal MSNs (Doyle *et al.*, 2008; Heiman *et al.*, 2008). In our study, we showed that RNA for microarray profiling can be obtained from live FACS purified neurons even at adult ages. Therefore, with the many fluorescent neuron transgenic lines available this technique is suitable to assess cell subtype genetic changes across the brain. Currently researchers continue to use FACS to profile gene expression in many cell subtypes in the nervous system (Cahoy *et al.*, 2008; Guez *et al.*, 2008; Marsh *et al.*, 2008).

In our FACS-array study, we were able to find a new set of genes that were differentially enriched in each MSN subtype. The most enriched gene we found in striatonigral MSNs was *Slc35d3*, an uncharacterized nucleotide sugar transporter. We further validated the striatonigral specificity of this gene by RT-PCR, double ISH, and a BAC transgenic, which displayed lack of colocalization with enkephalin protein and projections to striatonigral targets, the GPi and SN (Lobo *et al.*, 2006). Members of the Slc35 family transport sugar nucleotides to the ER and Golgi for protein glycosylation (Ishida and Kawakita, 2004) and Slc35 homologs in *Drosophila* (Goto *et al.*, 2001; Selva *et al.*, 2001) and *C. elegans* (Berninsone *et al.*, 2001) are rate-limiting regulators of protein glycosylation, controlling function of developmental molecules such as Notch, which is involved in cell fate decisions (Goto *et al.*, 2001; Selva *et al.*, 2001). Finally, functional insight into this gene is being evaluated by generation of a *Slc35d3* knockout, which displays motor deficits and striatal MSN developmental deficits (Lu *et al.*, 2008). Continued analyses of the *Slc35d3* knockout may provide a better understanding of striatonigral development and function.

Another enriched gene we found in this study was Zinc finger protein 521 (*Zfp521*) (Lobo *et al.*, 2006), a transcription factor that was originally identified as a retroviral-integration activated gene, which causes B-cell lymphomas (Justice *et al.*, 1994; Warming *et al.*, 2003). Additionally, *Zfp521* has been implicated to directly regulate early B-cell factor 1 (*Ebf1*) (Bond *et al.*, 2004; Hentges *et al.*, 2005; Warming *et al.*, 2003), a transcription factor necessary for B-cell differentiation (Lin and Grosschedl, 1995). Further analysis is needed to understand the precise role of *Zfp521* in striatonigral MSNs. However, we did focus on *Ebf1* due to its link with *Zfp521* and because *Ebf1* knockout mice exhibit deficits in striatal matrix development (Garel *et al.*, 1999). We also showed *Ebf1* to be enriched in striatonigral MSNs using RT-PCR and double ISH. Additionally, Ebf1 knockout mice that were crossed with striatonigral EGFP reporter mice display decreased EGFP neurons in the matrix compartments but not the patch compartments of dStr at postnatal day 20 (P20) but not P0 and a loss of EGFP terminals to the SN at both ages (Lobo *et al.*, 2006, 2008). These data imply that striatonigral MSNs in the matrix are specified at early postnatal ages but in the absence of *Ebf1* fail to form proper terminals to the SN and differentiate into mature striatonigral MSNs. Furthermore, we showed that in the absence of *Ebf1* there is a decrease in *Zfp521* (Lobo *et al.*, 2008) suggesting a link between these two molecules is not limited to

B-cell development but also includes striatonigral MSN development. There has been much research focused on molecular mechanisms in development of the patch and matrix compartments (Anderson *et al.*, 1997; Garel *et al.*, 1999; Mason *et al.*, 2005; Toresson and Campbell, 2001; Yun *et al.*, 2003) but virtually no information in known about the different molecular players in striatonigral versus striatopallidal MSN development. Out data provided the first glimpse of a potentially complex molecular program that orchestrates the differentiation of the striatonigral versus striatopallidal MSNs in the patch versus the matrix compartments. Nonetheless there is still little known about how the two MSNs develop. A more complex genetic profiling of the two MSNs at embryonic and early postnatal ages is key to identifying new molecules involved in differential development of these two neuronal subtypes.

A gene we found to be enriched in striatopallidal MSNs using this FACS method and then RT-PCR was G-protein couple receptor 6 (*Gpr6*) (Lobo *et al.*, 2007). We chose to evaluate it because GENSAT had made a *Gpr6*-EGFP BAC transgenic, which displayed colocalization of EGFP striatal MSNs with enkephalin protein and EGFP projections to the GPe but not GPi and SN (Lobo *et al.*, 2007; http://www.gensat.org) confirming the enrichment of *Gpr6* in striatopallidal MSNs. This was further validated with a *Gpr6* knockout mouse, in which *Gpr6* is replaced with a LacZ reporter gene. This mouse exhibited colocalization of LacZ with enkephalin protein and *Drd2*-EGFP but not *Drd1a*-EGFP when crossed with the two BAC transgenic mice (Lobo *et al.*, 2007). *Gpr6* has been identified as a Sphigosine-1-phosphate (S1P) intracellular receptor but can also constitutively activate G_s and $G_{i/o}$ signaling with an overall stimulation of G_s resulting in activation of adenylyl cyclase, increased levels of cAMP, and mobilization of Ca^{2+} stores (Ignatov *et al.*, 2003; Padmanabhan *et al.*, 2009; Uhlenbrock *et al.*, 2002). Our evaluation of the *Gpr6* knockout revealed that lack of *Gpr6* results in decreased striatal cAMP levels and enhanced instrumental conditioning for a sugar pellet (Lobo *et al.*, 2007). These data implicate the role for striatopallidal MSNs in instrumental learning and suggest *Gpr6* mediated cAMP signaling as an underlying molecular mechanism controlling this behavior in these neurons.

Thus far we have addressed the first large-scale microarray profiling of differential gene expression in striatonigral and striatopallidal MSNs using the FACS-array method. However, recently another group assessed genetic differences in these two neurons using a novel technique, TRAP, which involves expression of EGFP-tagged ribosomal protein L10a in specific cell types using BAC transgenics (bacTRAP) (Doyle *et al.*, 2008; Heiman *et al.*, 2008). This technique allows tagging of polysomes for immunoaffinity precipitation of translating mRNA in these different neuronal subtypes (*ibid.*). These researchers used *Drd1a*- and *Drd2*-bacTRAP mice to obtain mRNA from striatonigral and striatopallidal MSNs for microarray profiling (Heiman *et al.*, 2008). This study identified similar genes found in the FACS study including those previously addressed;

Slc35d3, *Zfp521*, *Ebf1*, and *Gpr6* (Lobo *et al.*, 2006, 2007) but also identified other novel sets of genes differentially enriched in the two MSNs. Furthermore, this study showed that application of S1P to *Drd1a*- and *Drd2*-EGFP mice resulted in increased dendritic Ca^{2+} levels in *Drd2*-EGFP striatopallidal MSNs but not *Drd1a*-EGFP striatonigral MSNs (Heiman *et al.*, 2008) which supports S1P-*Gpr6* signaling in striatopallidal MSN.

The bacTRAP study went beyond wild-type profiling to evaluate genetic changes in the two MSNs after acute and chronic cocaine treatments. This study further showed differential expression of genes known to be regulated in cocaine such as *FosB*, *Homer1*, *Vamp2*, *Kcnd2*, and *Zfp64* (Brakeman *et al.*, 1997; Hope *et al.*, 1992; McClung *et al.*, 2004). In particular, *FosB* mRNA levels are consistent with upregulation of FosB proteins in both cell types after acute cocaine exposure but continued expression of ΔFosB in striatonigral MSNs after chronic cocaine treatment (Lee *et al.*, 2006; Moratalla *et al.*, 1996; Nye *et al.*, 1995). Furthermore, this study revealed upregulation of GABA signaling genes in striatonigral MSNs after chronic cocaine and provided functional insight by showing increased $GABA_A$ receptor mediated miniature inhibitory postsynaptic currents (mIPSCs) in striatonigral MSNs after chronic cocaine treatment (Heiman *et al.*, 2008). Collectively these data from both MSN subtype gene expression studies emphasize the importance of genetic profiling of the two MSNs, which can lead to functional insights in the normal and diseased BG.

The BG field is now at a very promising stage where researchers can perform large-scale gene expression profiling of the two MSNs not just in the normal adult brain but also during development and in the diseased BG. We now have many techniques available to perform molecular profiling including FACS, laser capture, hand picking of neurons combined with fluorescent reporter mice or TRAP using the bacTRAP mice. However, there are other tools and methodologies researchers can perform to understand distinct molecular properties in these two MSNs. The remainder of this chapter will address current use of striatonigral and striatopallidal fluorescent reporter BAC transgenic mice to further assess molecular and signaling differences in these mice. This chapter will also address other types of profiling such as chromatin changes in the two MSNs and will discuss the use of cell type-specific knockout or overexpression models to fully understand the differential role certain molecules play in these two MSN subtypes.

IV. The Next Stage for Genetic Profiling of the Two MSNs

Thus far we have discussed gene expression profiling in the two MSNs however advanced techniques in neuroscience and genetics can help take the molecular profiling of these two neurons to the next level. We are able to isolate these two neurons using FACS but it would be advantageous to isolate and evaluate populations of neurons within the two MSNs. One such way would be to generate multiple fluorescent colored transgenic lines that can be crossed to the *Drd1a*- and *Drd2*-EGFP mice. For example, researchers can generate tdTomato or dsRed reporter lines for *calbindin* (a matrix marker) or mu opioid receptor (*muOR*, a patch marker) and they can be crossed to the *Drd1a*- and *Drd2*-EGFP mice for FACS purification and gene expression analysis of striatonigral versus striatopallidal MSNs in the patch versus matrix compartments. Our previous work implicated differential developmental processes at least for the striatonigral matrix neurons versus striatonigral patch neurons (Lobo *et al.*, 2008), therefore isolating the two MSNs in the patch versus matrix compartment during development would be valuable to gain insight into the molecular machinery governing development of these subclass of neurons. Additionally, this would be a useful technique to use with OCD spectrum models since research evaluating models of stereotypy, a hallmark phenotype of these disorders, reveals a molecular imbalance of the patch and matrix compartments (Canales and Graybiel, 2000; Graybiel and Rauch, 2000).

Another approach to evaluating subclasses of neurons within the two MSNs would involve immunostaining the two striatal MSNs with fluorescent antibodies then using FACS to purify the multiple neuronal populations (Fig. 4A). One group is already successfully FACS purifying fixed and immunostained adult striatal neuronal populations for gene expression profiling. They are utilizing a c-Fos lacZ reporter rat and immunostaining for b-gal and NeuN to isolate c-Fos+ neurons involved in context-specific locomotor sensitization to compare to cFos- control neurons (Guez *et al.*, 2008). They are successfully obtaining RNA from fixed neurons for gene expression profiling (*ibid.*) and this is the first time researchers are able to FACS purify adult fixed, immunostained neurons for genetic analyses. This novel technique can be used with the *Drd1a*- and *Drd2*-EGFP mice, to isolate subclasses within these two MSNs and would bypass the need for multiple transgenic reporter mice which can be timely to generate and breed.

The next stage in genetic profiling of the two MSNs would be to evaluate epigenetic differences in these two neuronal populations. The merge of epigenetics and neuroscience is in its nascent stage and researchers are now evaluating chromatin remodeling in striatum using ChIP in models of drug abuse and stress (Kumar *et al.*, 2005; Levine *et al.*, 2005; Pulipparacharuvil *et al.*, 2008; Renthal *et al.*, 2007, 2008, 2009; Russo *et al.*, 2009; Wilkinson *et al.*, 2009). These studies have revealed important insights into chromatin modifications on key genetic players known to be involved in chronic drug abuse and stress. However, these studies assessed chromatin modifications in whole striatal tissues, which include both MSN subtypes as well as interneurons and glia. Therefore, the next phase in evaluating epigenetic mechanisms in psychiatric and neurological disorders of the striatum would ideally involve partitioning the differential chromatin modifications occurring in the two MSN subtypes. However, this is not a trivial task since large amounts of material are necessary for ChIP experiments (Kumar *et al.*, 2005). Current advances in epigenetics may help to resolve these technical difficulties. Recently researchers have been able to evaluate epigenetic modifications in small cell numbers with as little as 100 cells (Dahl and Collas, 2007a,b, 2008; Dahl *et al.*, 2009; O'Neill *et al.*, 2006) and one of these studies involved a large genome-scale analyses using ChIP–chip on as few as 1000 cells (Dahl *et al.*, 2009). Therefore, it is plausible that one can combine FACS of *Drda1*- and *Drd2*-fluorescent reporter mice with these small cell number ChIP techniques to evaluate epigenetic differences in these two MSN subtypes (Fig. 4B). These are key experiments necessary for the field to advance and fully unravel the molecular disparities between the two MSN subtypes during development and in neuropsychiatric and neurological diseases.

V. Use of BAC Transgenic Fluorescent Reporter Mice to Evaluate Molecular Pathways in the Two MSNs

In the past 5 years the BG researchers have begun to exploit the *Drd1a*- and *Drd2*-EGFP BAC transgenic mice to evaluate electrophysiological differences and differential molecular and signaling pathways with the latter involving immunofluorescent techniques. The scope of this chapter is focused on molecular profiling so we will address information researchers have recently gathered on differential signaling. Nonetheless, it is important to note that many insights into differential electrophysiological properties including synaptic plasticity, DA modulation, cholinergic modulation, intrinsic excitability, glutamatergic activity, and GABA activity have been made using these mice (Ade *et al.*, 2008; Cepeda *et al.*, 2008; Day *et al.*, 2006; Ding *et al.*, 2008; Gertler *et al.*, 2008; Kreitzer and Malenka, 2007, 2008; Shen *et al.*, 2007, 2008; Taverna *et al.*, 2008; Wang *et al.*, 2006b).

Researchers are utilizing the *Drd1a*- and *Drd2*-EGFP BAC transgenic mice to evaluate molecular changes and signaling pathways in response to drugs of abuse, which promote DA release from the VTA, and also in response to DA depletion from the SN. Chronic treatment with cocaine or methylphenidate, both psychostimulants that promote excess DA in the synapse, followed by 1–2 days withdrawal caused an upregulation of the long lasting FosB isoform, ΔFosB in both striatonigral MSNs (NAc and dStr) and striatopallidal MSNs (NAc) for cocaine and only in striatonigral MSNs (NAc and dStr) for methylphenidate (Kim *et al.*, 2009; Lee *et al.*, 2006). Additionally, a longer withdrawal period (1 month) after cocaine resulted in ΔFosB induction only in striatonigral MSNs in the NAc suggesting the effect is transient in striatopallidal MSNs (Lee *et al.*, 2006).

These data are consistent with previous data showing the effects of cocaine and ΔFosB expression are mediated through striatonigral MSNs and not striatopallidal MSNs (Moratalla *et al.*, 1996; Nye *et al.*, 1995). Furthermore, these studies showed changes in dendritic spine density with increases occurring in striatonigral NAc (core and shell) MSNs and striatopallidal NAc (core only) MSNs after chronic methylphenidate and increase spine density in both MSNs in all striatal regions after chronic cocaine (Lee *et al.*, 2006). However, similar to ΔFosB expression after cocaine these spine changes disappear after 1 month withdrawal in striatopallidal MSNs but not striatonigral MSNs (Lee *et al.*, 2006) implicating ΔFosB mediation of spine density changes in the NAc after chronic cocaine (Kim *et al.*, 2009; Lee *et al.*, 2006).

Additionally, extracellular related kinase (ERK) signaling has been evaluated in the *Drd1a*- and *Drd2*-EGFP BAC in response to drugs of abuse, antipsychotics, and L-DOPA treatment in Parkinson's Disease models (Bertran-Gonzalez *et al.*, 2008, 2009; Borkvist *et al.*, 2008; Santini *et al.*, 2009). Acute and chronic cocaine activates phosphorylation of ERK in striatonigral MSNs in dStr and NAc and the acute treatment also increases phosphorylation of ERK targets, mitogen and stress activated kinase-1 (MSK-1) and histone H3 (Brami-Cherrier *et al.*, 2005), in the same neurons while chronic treatments only lead to increase in ERK phosphorylation in dStr (Bertran-Gonzalez *et al.*, 2008). This affect is not seen in the *Drd2*-EGFP striatopallidal MSNs (*ibid.*). However, an acute dose of haloperidol, a D2 receptor antagonist, leads to phosphorylation of ERK, MSK-1, and H3 in the striatopallidal MSNs (*ibid.*). A similar effect was seen for induction of immediate early genes (IEGs) c-Fos and Zif268 (*ibid.*), which are thought to be activated through ERK signaling (Brami-Cherrier *et al.*, 2005; Valjent *et al.*, 2000). IEG increase was predominantly in striatonigral MSNs after cocaine and predominantly in striatopallidal MSNs after haloperidol (*ibid.*), which is consistent with ERK activation of these IEGs. However, cocaine activated IEGs in a small population of striatopallidal MSNs and haloperidol activate IEGs in a small population of striatonigral MSNs suggesting a non-ERK mediated mechanism to activate IEGs in the opposite MSNs (*ibid.*). These data demonstrate differential means of evoking similar signaling cascades in the two MSNs. Interestingly, a higher number of p-H3 neurons compared to p-ERK neurons where observed in striatopallidal MSNs after haloperidol in contrast to cocaine which resulted in all p-H3 striatonigral MSNs colocalizing with p-ERK (*ibid.*). Further investigation revealed that the induction of p-H3 in striatopallidal MSNs does not require ERK signaling since this induction was not affected when ERK signaling was blocked (Bertran-Gonzalez *et al.*, 2009). However, the p-H3 induction does require G_{olf} signaling likely through Adora2a activation and this signaling also involves PKA phosphorylation of DARPP-32 (*ibid.*). Similar to the cocaine induction of p-ERK, p-MSK-1, and p-H3 in striatonigral MSNs (Bertran-Gonzalez *et al.*, 2008), L-DOPA treatment in 6-hydroxydopamine (6-OHDA) lesioned (a model for

Parkinson's Disease) *Drd1a*- and *Drd2*-EGFP mice also induced ERK signaling in striatonigral MSNs (Santini *et al.*, 2009). Evidence suggests increased ERK signaling activity in L-DOPA induced dyskinesias, a deleterious motor side effect of L-DOPA treatment in Parkinson's Disease (Santini *et al.*, 2007; Schuster *et al.*, 2008) and these data implicate a role for striatonigral increased ERK signaling in L-DOPA induced dyskinesias. Finally, ERK phosphorylation has been evaluated in the *Drd1a*- and *Drd2*-EGFP mice in morphine sensitization. Context-specific induction (induction in the drug paired cage but not home cage) of p-ERK was observed after withdrawal primarily in the striatonigral MSNs and a small number of striatopallidal MSNs and this induction can be blocked with a D1 receptor antagonist, indicating a role for ERK signaling in striatonigral neurons during reexposure to the drug context (Borgkvist *et al.*, 2008). These studies have revealed important insights into differential activation of signaling pathways in the two MSNs in response to DA modulation. Without the use of the *Drd1a*- and *Drd2*-EGFP BAC mice these studies would have been much more difficult requiring double ISH/immunohistochemistry experiments or would have been impossible in the case of some of the triple-immunolabeling experiments. The BG field is at an extraordinary time point where studies evaluating differential signaling pathways, other protein differences, morphological and dendritic spine differences, genetic differences, and physiological differences can easily being employed utilizing these BAC transgenic mice. These mice have opened the gates toward a more comprehensive understanding of the differential roles these two MSNs play in the normal, diseased, and developing brain.

VI. Cell Type-Specific Overexpression and Knockout Models

The final stage toward understanding differential molecular mechanisms in the two MSNs involves conditional mouse models where specific genes can be expressed or silenced in either MSN subtype. Researchers are beginning to use knock-in or BAC transgenic mice that express Cre-recombinase under *Drd1a* or *Drd2* promoters (Gong *et al.*, 2007; Heusner *et al.*, 2008; Lemberger *et al.*, 2007) and are crossing these lines to conditional mouse models. These models make use of the Cre/LoxP system in which DNA floxed by two Lox P cites can be inverted or excised, depending on the direction of the Lox P cites, by Cre-recombinase. This system in most commonly being used with conditional floxed knockout mice. However, other genetic manipulations can be used with the Cre driver lines including conditional overexpression of siRNA viral systems. Adeno-associated virus (AAV) or herpes simplex virus (HSV) with gene constructs that include a LoxP-Stop-LoxP sequence prior to the gene (Kuhlman and Huang, 2008) or inverted double floxed genes (Atasoy *et al.*, 2008; Tsai *et al.*, 2009) (with both

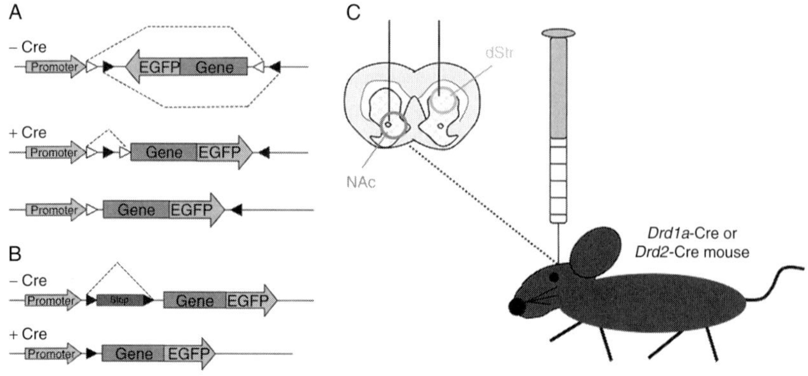

FIG. 5. Conditional AAV or HSV can be infused into striatal areas in *Drd1a*-Cre or *Drd2*-Cre mice to express specific genes in each MSN subtype. (A) A conditional viral vector strategy consisting of the gene in reverse with a fluorescent tag, floxed by two sets of incompatible lox sites. In the presence of Cre, an event occurs at one lox set resulting in an inversion of the gene. This is followed by a second event with removes two of the incompatible lox sites. The final result is the gene in the correct transcriptional orientation floxed by two incompatible lox sites, with no further events. (B) An alternative viral vector strategy in which a stop sequence floxed by two loxP sites precedes a gene with a fluorescent tag. The gene is transcriptionally inactive until Cre is present and can excise the stop sequence. (C) Conditional viruses can be stereotaxically infused into *Drd1a*-Cre or *Drd2*-Cre transgenic mice to allow for expression of a specific gene in either MSN subtype. Furthermore, this can be done at adult ages to avoid developmental consequences and the virus can be infused into specific striatal regions, for example, NAc (blue) or dStr (pink).

constructs allowing gene transcription in the presence of Cre) can be stereotaxically injected into the dStr or NAc of *Drd1a*- Cre or *Drd2*-Cre transgenic mice (Fig. 5). Conditional viral targeting strategies can allow for temporal, striatal region (dStr vs. NAc), and MSN subtype specificity while only the latter can be achieved with conditional knockout and overexpression mouse models. Along with the *Drd1a*-Cre and *Drd2*-Cre lines, researchers are generating transgenic mice, which overexpress specific genes under *Drd1a* or *Drd2* promoters (Drago *et al.*, 1998; Gantois *et al.*, 2007; Heusner and Palmiter, 2005; Sano *et al.*, 2003). However, despite these recent advances in Cre driver lines and overexpression models using, BACs and knock-ins, the first conditional striatonigral and striatopallidal models were the NSE-tTa (tetracycline transactivator) lines. In this system transgenic mice expressing a gene of interest driven under Tet-Op (tetracycline promoter) can be crossed with NSE-tTa lines and in the absence of doxycycline (a tetracycline derivative) tTa will activate Tet-Op to induce gene expression (Furth *et al.*, 1994). Gene expression can remain off throughout development while the mouse is supplemented with doxycycline, then later turned on at adult ages. These types of models described above are important for understanding the functional role of specific genes in each MSN subtype. The subsequent sections will discuss functional insights researchers have made into the two MSNs by using these cell subtype-specific models.

VII. Cell Type-Specific Mouse Models Evaluating Drug and Natural Reward

Nestler and colleagues generated two NSE-tTa lines, termed11A and 11B, which express tTa in either striatonigral or striatopallidal MSNs, respectively (Chen *et al.*, 1998; Kelz *et al.*, 1999; Werme *et al.*, 2002). Line 11A mice crossed with a Tet-Op ΔFosB line show increased response to the rewarding and locomotor effects of cocaine (Kelz *et al.*, 1999), which is consistent with ΔFosB induction in striatonigral MSNs (Moratalla *et al.*, 1996; Nye *et al.*, 1995). Furthermore, these same mice display increased reward to morphine while the 11B × TetOp ΔFosB mice show no increase in morphine reward further supporting the role of ΔFosB induction in striatonigral MSNs as an important molecular player in the rewarding properties of drugs of abuse (Zachariou *et al.*, 2006). This phenomenon is also observed in other reward behaviors in particular wheel running, a known natural reward (Belke, 1997; Iversen, 1993; Lett *et al.*, 2000). Line 11A × Tet-Op deltaFosB mice display increased wheel running behavior, in contrast to 11B × Tet-Op ΔFosB mice which exhibit decreased wheel running compared to littermate controls (Werme *et al.*, 2002). This further supports a role for ΔFosB in striatonigral MSNs in reward behaviors and implicates induction of ΔFosB in striatopallidal MSNs as having a negative consequence on rewarding stimuli.

Other groups have evaluated striatonigral function in drug reward. One group generated a knock-in mouse expressing a mutated form of the NMDAR1 subunit (the mutation is in the pore of the subunit and reduces calcium influx thereby decreasing NMDA signaling) driven under the *Drd1a* promoter (Heusner and Palmiter, 2005). They show that reduced NMDA signaling in the striatonigral MSNs results in a reduction to the sensitization of chronic cocaine treatment and a lack of cocaine conditioned place preference at 20 mg/kg but not higher doses of 30 mg/kg suggesting these mice are less sensitive to the rewarding effects of cocaine (*ibid.*) implicating a role for striatonigral glutamatergic NMDA signaling in cocaine reward.

More recently researchers have assessed the role of the glucocorticoid receptor in striatonigral MSNs in addiction. Floxed *Nr3c1* (the glucocorticoid receptor gene) where crossed with a *Drd1a*-Cre BAC transgenic line to generate mice lacking the *Nr3c1* gene in striatonigral MSNs (Ambroggi *et al.*, 2009). Previous research revealed that loss of *Nr3c1* throughout the brain results in decreased cocaine self-administration (Deroche-Gamonet *et al.*, 2003). *Drd1a* specific *Nr3c1* knockouts display a similar phenotype and these mice show a decrease in VTA DA neuron firing (Ambroggi *et al.*, 2009) implicating a role for striatonigral MSNs in the motivational properties of cocaine reward and in regulating VTA activity during reward behaviors. However, it is plausible that cortical neurons expressing *Drd1a* may play a role in the phenotypes observed. Although *Drd1a* and *Drd2* are most abundantly enriched in striatum, they are also expressed in other brain regions including cortex and hippocampus for *Drd1a* and cortex and SN for *Drd2* (Allen Brain Atlas—http://www.brain-map.org). This is one of the limitations of expressing Cre or other genes under *Drd1a* or *Drd2* for researchers wanting to study striatonigral and striatopallidal MSNs but these limitations can be overcome with the use of conditional viruses described above.

Another group evaluated drug behavior in the two MSNs by generating a conditional striatopallidal MSN ablation model. They used an *Adora2a*-Cre BAC transgenic line, which they crossed to an inducible diphtheria toxin mouse line. Ablation of striatopallidal MSNs in the NAc produced an increased response to amphetamine reward implicating a role for striatopallidal MSNs in normal attenuation of drug reward (Durieux *et al.*, 2009). Further assessment of striatonigral MSN ablation models, models in which striatonigral or striatopallidal MSN activity is modulated, and models manipulating specific genes in each populations may help further elucidate the roles the two MSNs play in addictive behaviors.

VIII. Cell Type-Specific Mouse Models Evaluating Motor Behaviors

Researchers have also evaluated BG motor phenotypes using striatonigral and striatopallidal specific mouse models. Groups have generated ablation models in which *Drd1a* or *Drd2* promoters drive genes that result in cell ablation or the *Adora2a*-Cre specific ablation model described above (Drago *et al.*, 1998; Durieux *et al.*, 2009; Gantois *et al.*, 2007; Sano *et al.*, 2003). In these models, MSNs are ablated during development for the striatonigral models and at adult ages for the striatopallidal model and some of these models support the traditional BG direct versus indirect model where the direct pathway promotes movement and the indirect pathway inhibits movement (Drago *et al.*, 1998; Durieux *et al.*, 2009; Sano *et al.*, 2003). However, one model, in which striatonigral MSNs are ablated around the second postnatal week, displays hyperactive behavior (Gantois *et al.*, 2007). These data suggest a more complex BG network than the proposed model. Nonetheless, more detailed research including a model in which striatonigral MSNs are ablated at adult ages is necessary to fully understand the role of these two MSNs in BG motor behaviors.

Another group knocked out GAD67, the enzyme responsible for GABA synthesis, in striatonigral MSNs using a Drd1a-Cre knock-in mouse and then evaluated motor behaviors (Heusner *et al.*, 2008) and showed that loss of GABA release (which is the main neurotransmitter released from MSNs) from striatonigral MSNs resulted in mild motor deficits (*ibid.*). These cell type-specific genetic models are helping scientists unravel the precise role the two MSNs play in motor function. These genetic models can further define and asses the traditional BG model of the direct pathway promoting movement and the indirect pathway inhibiting movement to allow for a comprehensive understanding of the role of these two MSNs in motor function and movement disorders.

IX. Cell Type-Specific Models Evaluating Protein Modifications

Researchers are also assessing posttranslational protein modifications in the two MSNs using cell type-specific models. In particular, researchers are studying DARPP-32 (DA and cAMP regulated phosphoprotein of 32 kDa), an important regulator of signaling in striatal MSNs. Phosphorylation of DARPP-32 at threonine 34 (T34) by PKA results in inhibition of protein phosphatase-1 (PP-1), whereas phosphorylation at threonine 75 (T75) by cdk5 causes inhibition of PKA signaling and PP-1 is no longer inhibited (Svenningsson *et al.*, 2005). Differential inhibition of either PKA or PP-1 by DARPP-32 can result in differential modulation of downstream molecules that play important roles in transcription, physiology, behavior, and pharmacological stimuli (*ibid.*). To assess DARPP-32 signaling in the two MSNs, Greengard and colleagues have generated BAC transgenic models in which they are able to evaluate phosphorylation of DARPP-32 in either striatopallidal or striatonigral MSNs by expressing tagged versions of Darpp-32 using *Drd1a* or *Drd2* BACs (Bateup *et al.*, 2008). Using this technique, they were able to immunoprecipitate Darpp-32 from each MSN subtype and then evaluate phosphorylation changes in response to cocaine and haloperidol. They show that acute cocaine treatment increases T34 phosphorylation in striatonigral MSNs with a decrease in striatopallidal MSNs and T75 phosphorylation is up in striatopallidal MSNs while down in the striatonigral MSNs. Additionally, haloperidol (a D2 antagonist) treatment induced phosphorylation at T34 in striatopallidal MSNs with no change in striatonigral MSNs. This unique cell type-specific analysis was important for evaluating the differential phosphorylation of Darpp-32 in striatonigral and striatonigral MSNs and this type of analysis can be extended to evaluate posttranslational modifications in other proteins within the two MSN subtypes.

X. Conclusions

The BG field has come along way in the past 20 years and has made extraordinary progress in recent years with the availability of techniques to isolate these two neurons and profile their molecular identities. BG researchers are now equipped with the most sophisticated tools to comprehensively evaluate molecular properties and their functional consequences in the two MSN subtypes. We now have specialized tools including FACS of fluorescent reporter mice and bacTRAP to evaluate large-scale gene expression differences in the two MSNs and will likely be able to apply these techniques to assess large-scale chromatin differences in the two MSNs. We are also equipped with cell type-specific models using Cre driver lines and conditional floxed mouse models and viruses to evaluate distinct genetic functions in the two MSNs. These methodologies are imperative for understanding distinct properties in these two neurons during development and diseased states. The field is poised at a thrilling time point where we are well outfitted with the essential tools to truly dissect distinct molecular mechanisms underlying functional output in these two MSNs. This is an extraordinary time to be a researcher in this field and one can predict that armed with this display of progressive techniques the next few years will yield exciting insights into these two MSNs.

References

Ade, K. K., Janssen, M. J., Ortinski, P. I., and Vicini, S. (2008). Differential tonic GABA conductances in striatal medium spiny neurons. *J. Neurosci.* **28,** 1185–1197.

Albin, R. L., Young, A. B., and Penney, J. B. (1989). The functional anatomy of basal ganglia disorders. *Trends Neurosci.* **12,** 366–375.

Ambroggi, F., Turiault, M., Milet, A., Deroche-Gamonet, V., Parnaudeau, S., Balado, E., Barik, J., van der Veen, R., Maroteaux, G., Lemberger, T., Schutz, G., Lazar, M., *et al.* (2009). Stress and addiction: Glucocorticoid receptor in dopaminoceptive neurons facilitates cocaine seeking. *Nat. Neurosci.* **12,** 247–249.

Anderson, S. A., Qiu, M., Bulfone, A., Eisenstat, D. D., Meneses, J., Pedersen, R., and Rubenstein, J. L. (1997). Mutations of the homeobox genes Dlx-1 and Dlx-2 disrupt the striatal subventricular zone and differentiation of late born striatal neurons. *Neuron* **19,** 27–37.

Arlotta, P., Molyneaux, B. J., Chen, J., Inoue, J., Kominami, R., and Macklis, J. D. (2005). Neuronal subtype-specific genes that control corticospinal motor neuron development *in vivo*. *Neuron* **45,** 207–221.

Atasoy, D., Aponte, Y., Su, H. H., and Sternson, S. M. (2008). A FLEX switch targets Channelrhodopsin-2 to multiple cell types for imaging and long-range circuit mapping. *J. Neurosci.* **28,** 7025–7030.

Bateup, H. S., Svenningsson, P., Kuroiwa, M., Gong, S., Nishi, A., Heintz, N., and Greengard, P. (2008). Cell type-specific regulation of DARPP-32 phosphorylation by psychostimulant and antipsychotic drugs. *Nat. Neurosci.* **11,** 932–939.

Belin, D., Jonkman, S., Dickinson, A., Robbins, T. W., and Everitt, B. J. (2008). Parallel and interactive learning processes within the basal ganglia: Relevance for the understanding of addiction. *Behav. Brain Res.* **199,** 89–102.

Belke, T. W. (1997). Running and responding reinforced by the opportunity to run: Effect of reinforcer duration. *J. Exp. Anal. Behav.* **67,** 337–351.

Bernard, V., Normand, E., and Bloch, B. (1992). Phenotypical characterization of the rat striatal neurons expressing muscarinic receptor genes. *J. Neurosci.* **12,** 3591–3600.

Berninsone, P., Hwang, H. Y., Zemtseva, I., Horvitz, H. R., and Hirschberg, C. B. (2001). SQV-7, a protein involved in *Caenorhabditis elegans* epithelial invagination and early embryogenesis, transports UDP-glucuronic acid, UDP-N-acetylgalactosamine, and UDP-galactose. *Proc. Natl. Acad. Sci. USA* **98,** 3738–3743.

Berridge, K. C., and Robinson, T. E. (1998). What is the role of dopamine in reward: Hedonic impact, reward learning, or incentive salience? *Brain Res. Brain Res. Rev.* **28,** 309–369.

Bertran-Gonzalez, J., Bosch, C., Maroteaux, M., Matamales, M., Herve, D., Valjent, E., and Girault, J. A. (2008). Opposing patterns of signaling activation in dopamine D1 and D2 receptor-expressing striatal neurons in response to cocaine and haloperidol. *J. Neurosci.* **28,** 5671–5685.

Bertran-Gonzalez, J., Hakansson, K., Borgkvist, A., Irinopoulou, T., Brami-Cherrier, K., Usiello, A., Greengard, P., Herve, D., Girault, J. A., Valjent, E., and Fisone, G. (2009). Histone H3 phosphorylation is under the opposite tonic control of dopamine D2 and adenosine A2A receptors in striatopallidal neurons. *Neuropsychopharmacology.* **34,** 1710–1720.

Bond, H. M., Mesuraca, M., Carbone, E., Bonelli, P., Agosti, V., Amodio, N., De Rosa, G., Di Nicola, M., Gianni, A. M., Moore, M. A., Hata, A., Grieco, M., *et al.* (2004). Early hematopoietic zinc finger protein (EHZF), the human homolog to mouse Evi3, is highly expressed in primitive human hematopoietic cells. *Blood* **103,** 2062–2070.

Borgkvist, A., Valjent, E., Santini, E., Herve, D., Girault, J. A., and Fisone, G. (2008). Delayed, context- and dopamine D1 receptor-dependent activation of ERK in morphine-sensitized mice. *Neuropharmacology* **55,** 230–237.

Brakeman, P. R., Lanahan, A. A., O'Brien, R., Roche, K., Barnes, C. A., Huganir, R. L., and Worley, P. F. (1997). Homer: A protein that selectively binds metabotropic glutamate receptors. *Nature* **386,** 284–288.

Brami-Cherrier, K., Valjent, E., Herve, D., Darragh, J., Corvol, J. C., Pages, C., Arthur, S. J., Girault, J. A., and Caboche, J. (2005). Parsing molecular and behavioral effects of cocaine in mitogen- and stress-activated protein kinase-1-deficient mice. *J. Neurosci.* **25,** 11444–11454.

Brownstein, M. J., Mroz, E. A., Tappaz, M. L., and Leeman, S. E. (1977). On the origin of substance P and glutamic acid decarboxylase (GAD) in the substantia nigra. *Brain Res.* **135,** 315–323.

Buckley, N. J., Bonner, T. I., and Brann, M. R. (1988). Localization of a family of muscarinic receptor mRNAs in rat brain. *J. Neurosci.* **8,** 4646–4652.

Cahoy, J. D., Emery, B., Kaushal, A., Foo, L. C., Zamanian, J. L., Christopherson, K. S., Xing, Y., Lubischer, J. L., Krieg, P. A., Krupenko, S. A., Thompson, W. J., and Barres, B. A. (2008). A transcriptome database for astrocytes, neurons, and oligodendrocytes: A new resource for understanding brain development and function. *J. Neurosci.* **28,** 264–278.

Calabresi, P., Centonze, D., Gubellini, P., Pisani, A., and Bernardi, G. (2000). Acetylcholine-mediated modulation of striatal function. *Trends Neurosci.* **23,** 120–126.

Canales, J. J., and Graybiel, A. M. (2000). A measure of striatal function predicts motor stereotypy. *Nat. Neurosci.* **3,** 377–383.

Cardinal, R. N., Parkinson, J. A., Hall, J., and Everitt, B. J. (2002). Emotion and motivation: The role of the amygdala, ventral striatum, and prefrontal cortex. *Neurosci. Biobehav. Rev.* **26,** 321–352.

Cepeda, C., Andre, V. M., Yamazaki, I., Wu, N., Kleiman-Weiner, M., and Levine, M. S. (2008). Differential electrophysiological properties of dopamine D1 and D2 receptor-containing striatal medium-sized spiny neurons. *Eur. J. Neurosci.* **27,** 671–682.

Chao, J., and Nestler, E. J. (2004). Molecular neurobiology of drug addiction. *Annu. Rev. Med.* **55,** 113–132.

Chen, J., Kelz, M. B., Zeng, G., Sakai, N., Steffen, C., Shockett, P. E., Picciotto, M. R., Duman, R. S., and Nestler, E. J. (1998). Transgenic animals with inducible, targeted gene expression in brain. *Mol. Pharmacol.* **54,** 495–503.

Coppola, G., and Geschwind, D. H. (2006). Microarrays and the microscope: Balancing throughput with resolution. *J. Physiol.* **575,** 353–359.

Creese, I., Sibley, D. R., Hamblin, M. W., and Leff, S. E. (1983). The classification of dopamine receptors: Relationship to radioligand binding. *Annu. Rev. Neurosci.* **6,** 43–71.

Cuello, A. C., and Paxinos, G. (1978). Evidence for a long Leu-enkephalin striatopallidal pathway in rat brain. *Nature* **271,** 178–180.

Dahl, J. A., and Collas, P. (2007a). A quick and quantitative chromatin immunoprecipitation assay for small cell samples. *Front. Biosci.* **12,** 4925–4931.

Dahl, J. A., and Collas, P. (2007b). Q2ChIP, a quick and quantitative chromatin immunoprecipitation assay, unravels epigenetic dynamics of developmentally regulated genes in human carcinoma cells. *Stem. Cells* **25,** 1037–1046.

Dahl, J. A., and Collas, P. (2008). MicroChIP—A rapid micro chromatin immunoprecipitation assay for small cell samples and biopsies. *Nucleic. Acids. Res.* **36,** e15.

Dahl, J. A., Reiner, A. H., and Collas, P. (2009). Fast genomic muChIP–chip from 1,000 cells. *Genome. Biol.* **10,** R13.

Day, M., Wang, Z., Ding, J., An, X., Ingham, C. A., Shering, A. F., Wokosin, D., Ilijic, E., Sun, Z., Sampson, A. R., Mugnaini, E., Deutch, A. Y., *et al.* (2006). Selective elimination of glutamatergic synapses on striatopallidal neurons in Parkinson disease models. *Nat. Neurosci.* **9,** 251–259.

Del Fiacco, M., Paxinos, G., and Cuello, A. C. (1982). Neostriatal enkephalin-immunoreactive neurones project to the globus pallidus. *Brain Res.* **231,** 1–17.

DeLong, M. R., and Wichmann, T. (2007). Circuits and circuit disorders of the basal ganglia. *Arch. Neurol.* **64,** 20–24.

Deroche-Gamonet, V., Sillaber, I., Aouizerate, B., Izawa, R., Jaber, M., Ghozland, S., Kellendonk, C., Le Moal, M., Spanagel, R., Schutz, G., Tronche, F., and Piazza, P. V. (2003). The glucocorticoid receptor as a potential target to reduce cocaine abuse. *J. Neurosci.* **23,** 4785–4790.

Ding, J., Peterson, J. D., and Surmeier, D. J. (2008). Corticostriatal and thalamostriatal synapses have distinctive properties. *J. Neurosci.* **28,** 6483–6492.

Doyle, J. P., Dougherty, J. D., Heiman, M., Schmidt, E. F., Stevens, T. R., Ma, G., Bupp, S., Shrestha, P., Shah, R. D., Doughty, M. L., Gong, S., Greengard, P., *et al.* (2008). Application of a translational profiling approach for the comparative analysis of CNS cell types. *Cell* **135,** 749–762.

Drago, J., Padungchaichot, P., Wong, J. Y., Lawrence, A. J., McManus, J. F., Sumarsono, S. H., Natoli, A. L., Lakso, M., Wreford, N., Westphal, H., Kola, I., and Finkelstein, D. I. (1998). Targeted expression of a toxin gene to D1 dopamine receptor neurons by cre-mediated site-specific recombination. *J. Neurosci.* **18,** 9845–9857.

Durieux, P. F., Bearzatto, B., Guiducci, S., Buch, T., Waisman, A., Zoli, M., Schiffmann, S. N., and de Kerchove d'Exaerde, A. (2009). D2R striatopallidal neurons inhibit both locomotor and drug reward processes. *Nat. Neurosci.* **12,** 393–395.

Eberwine, J., Yeh, H., Miyashiro, K., Cao, Y., Nair, S., Finnell, R., Zettel, M., and Coleman, P. (1992). Analysis of gene expression in single live neurons. *Proc. Natl. Acad. Sci. USA* **89,** 3010–3014.

Fishell, G., and van der Kooy, D. (1987). Pattern formation in the striatum: Developmental changes in the distribution of striatonigral neurons. *J. Neurosci.* **7,** 1969–1978.

Fishell, G., and van der Kooy, D. (1989). Pattern formation in the striatum: Developmental changes in the distribution of striatonigral projections. *Brain Res. Dev. Brain Res.* **45,** 239–255.

Fishell, G., and van der Kooy, D. (1991). Pattern formation in the striatum: Neurons with early projections to the substantia nigra survive the cell death period. *J. Comp. Neurol.* **312,** 33–42.

Foehring, R. C., Mermelstein, P. G., Song, W. J., Ulrich, S., and Surmeier, D. J. (2000). Unique properties of R-type calcium currents in neocortical and neostriatal neurons. *J. Neurophysiol.* **84,** 2225–2236.

Forn, J., Krueger, B. K., and Greengard, P. (1974). Adenosine 3′,5′-monophosphate content in rat caudate nucleus: Demonstration of dopaminergic and adrenergic receptors. *Science* **186,** 1118–1120.

Furth, P. A., St Onge, L., Boger, H., Gruss, P., Gossen, M., Kistner, A., Bujard, H., and Hennighausen, L. (1994). Temporal control of gene expression in transgenic mice by a tetracycline-responsive promoter. *Proc. Natl. Acad. Sci. USA* **91,** 9302–9306.

Gantois, I., Fang, K., Jiang, L., Babovic, D., Lawrence, A. J., Ferreri, V., Teper, Y., Jupp, B., Ziebell, J., Morganti-Kossmann, C. M., O'Brien, T. J., Nally, R., *et al.* (2007). Ablation of D1 dopamine receptor-expressing cells generates mice with seizures, dystonia, hyperactivity, and impaired oral behavior. *Proc. Natl. Acad. Sci. USA* **104,** 4182–4187.

Garel, S., Marin, F., Grosschedl, R., and Charnay, P. (1999). Ebf1 controls early cell differentiation in the embryonic striatum. *Development* **126,** 5285–5294.

Gerfen, C. R. (1984). The neostriatal mosaic: Compartmentalization of corticostriatal input and striatonigral output systems. *Nature* **311,** 461–464.

Gerfen, C. R. (1992). The neostriatal mosaic: Multiple levels of compartmental organization in the basal ganglia. *Annu. Rev. Neurosci.* **15,** 285–320.

Gerfen, C. R., and Young, W. S. 3rd. (1988). Distribution of striatonigral and striatopallidal peptidergic neurons in both patch and matrix compartments: An *in situ* hybridization histochemistry and fluorescent retrograde tracing study. *Brain Res.* **460,** 161–167.

Gerfen, C. R., Engber, T. M., Mahan, L. C., Susel, Z., Chase, T. N., Monsma, F. J. Jr., and Sibley, D. R. (1990). D1 and D2 dopamine receptor-regulated gene expression of striatonigral and striatopallidal neurons. *Science* **250,** 1429–1432.

Gertler, T. S., Chan, C. S., and Surmeier, D. J. (2008). Dichotomous anatomical properties of adult striatal medium spiny neurons. *J. Neurosci.* **28,** 10814–10824.

Gong, S., Doughty, M., Harbaugh, C. R., Cummins, A., Hatten, M. E., Heintz, N., and Gerfen, C. R. (2007). Targeting Cre recombinase to specific neuron populations with bacterial artificial chromosome constructs. *J. Neurosci.* **27,** 9817–9823.

Gong, S., Yang, X. W., Li, C., and Heintz, N. (2002). Highly efficient modification of bacterial artificial chromosomes (BACs) using novel shuttle vectors containing the R6Kgamma origin of replication. *Genome. Res.* **12,** 1992–1998.

Gong, S., Zheng, C., Doughty, M. L., Losos, K., Didkovsky, N., Schambra, U. B., Nowak, N. J., Joyner, A., Leblanc, G., Hatten, M. E., and Heintz, N. (2003). A gene expression atlas of the central nervous system based on bacterial artificial chromosomes. *Nature* **425,** 917–925.

Goto, S., Taniguchi, M., Muraoka, M., Toyoda, H., Sado, Y., Kawakita, M., and Hayashi, S. (2001). UDP-sugar transporter implicated in glycosylation and processing of Notch. *Nat. Cell Biol.* **3,** 816–822.

Graybiel, A. M. (2000). The basal ganglia. *Curr. Biol.* **10,** R509–R511.

Graybiel, A. M. (2005). The basal ganglia: Learning new tricks and loving it. *Curr. Opin. Neurobiol.* **15,** 638–644.

Graybiel, A. M., and Rauch, S. L. (2000). Toward a neurobiology of obsessive-compulsive disorder. *Neuron* **28,** 343–347.

Guez, D. H., Harvey, B. K., Piciotto, M. R., and Hope, B. T. (2008). Fluorescence activated cell sorting: A novel method to study neurons selectively activated during context-specific cocaine sensitization. *Soc. Neurosci.* **2008,** 367.

Heiman, M., Schaefer, A., Gong, S., Peterson, J. D., Day, M., Ramsey, K. E., Suarez-Farinas, M., Schwarz, C., Stephan, D. A., Surmeier, D. J., Greengard, P., and Heintz, N. (2008). A translational profiling approach for the molecular characterization of CNS cell types. *Cell* **135,** 738–748.

Heintz, N. (2001). BAC to the future: The use of bac transgenic mice for neuroscience research. *Nat. Rev. Neurosci.* **2,** 861–870.

Hentges, K. E., Weiser, K. C., Schountz, T., Woodward, L. S., Morse, H. C., and Justice, M. J. (2005). Evi3, a zinc-finger protein related to EBFAZ, regulates EBF activity in B-cell leukemia. *Oncogene* **24,** 1220–1230.

Herrero, M. T., Barcia, C., and Navarro, J. M. (2002). Functional anatomy of thalamus and basal ganglia. *Childs Nerv. Syst.* **18,** 386–404.

Heusner, C. L., and Palmiter, R. D. (2005). Expression of mutant NMDA receptors in dopamine D1 receptor-containing cells prevents cocaine sensitization and decreases cocaine preference. *J. Neurosci.* **25,** 6651–6657.

Heusner, C. L., Beutler, L. R., Houser, C. R., and Palmiter, R. D. (2008). Deletion of GAD67 in dopamine receptor-1 expressing cells causes specific motor deficits. *Genesis* **46,** 357–367.

Hong, J. S., Yang, H. Y., and Costa, E. (1977a). On the location of methionine enkephalin neurons in rat striatum. *Neuropharmacology* **16,** 451–453.

Hong, J. S., Yang, H. Y., Racagni, G., and Costa, E. (1977b). Projections of substance P containing neurons from neostriatum to substantia nigra. *Brain Res.* **122,** 541–544.

Hope, B., Kosofsky, B., Hyman, S. E., and Nestler, E. J. (1992). Regulation of immediate early gene expression and AP-1 binding in the rat nucleus accumbens by chronic cocaine. *Proc. Natl. Acad. Sci. USA* **89,** 5764–5768.

Hyman, S. E., Malenka, R. C., and Nestler, E. J. (2006). Neural mechanisms of addiction: The role of reward-related learning and memory. *Annu. Rev. Neurosci.* **29,** 565–598.

Ignatov, A., Lintzel, J., Kreienkamp, H. J., and Schaller, H. C. (2003). Sphingosine-1-phosphate is a high-affinity ligand for the G protein-coupled receptor GPR6 from mouse and induces intracellular Ca2+ release by activating the sphingosine-kinase pathway. *Biochem. Biophys. Res. Commun.* **311,** 329–336.

Ince, E., Ciliax, B. J., and Levey, A. I. (1997). Differential expression of D1 and D2 dopamine and m4 muscarinic acetylcholine receptor proteins in identified striatonigral neurons. *Synapse* **27,** 357–366.

Ishida, N., and Kawakita, M. (2004). Molecular physiology and pathology of the nucleotide sugar transporter family (SLC35). *Pflugers Arch.* **447,** 768–775.

Iversen, I. H. (1993). Techniques for establishing schedules with wheel running as reinforcement in rats. *J. Exp. Anal. Behav.* **60,** 219–238.

Justice, M. J., Morse, H. C. 3rd., Jenkins, N. A., and Copeland, N. G. (1994). Identification of Evi-3, a novel common site of retroviral integration in mouse AKXD B-cell lymphomas. *J. Virol.* **68,** 1293–1300.

Kamme, F., Salunga, R., Yu, J., Tran, D. T., Zhu, J., Luo, L., Bittner, A., Guo, H. Q., Miller, N., Wan, J., and Erlander, M. (2003). Single-cell microarray analysis in hippocampus CA1: Demonstration and validation of cellular heterogeneity. *J. Neurosci.* **23,** 3607–3615.

Kamme, F., Zhu, J., Luo, L., Yu, J., Tran, D. T., Meurers, B., Bittner, A., Westlund, K., Carlton, S., and Wan, J. (2004). Single-cell laser-capture microdissection and RNA amplification. *Methods Mol. Med.* **99,** 215–223.

Kanazawa, I., Emson, P. C., and Cuello, A. C. (1977). Evidence for the existence of substance P-containing fibres in striato-nigral and pallido-nigral pathways in rat brain. *Brain Res.* **119,** 447–453.

Kebabian, J. W., and Calne, D. B. (1979). Multiple receptors for dopamine. *Nature* **277,** 93–96.

Kelz, M. B., Chen, J., Carlezon, W. A. Jr., Whisler, K., Gilden, L., Beckmann, A. M., Steffen, C., Zhang, Y. J., Marotti, L., Self, D. W., Tkatch, T., Baranauskas, G., *et al.* (1999). Expression of the transcription factor deltaFosB in the brain controls sensitivity to cocaine. *Nature* **401,** 272–276.

Kim, Y., Teylan, M. A., Baron, M., Sands, A., Nairn, A. C., and Greengard, P. (2009). Methylphenidate-induced dendritic spine formation and DeltaFosB expression in nucleus accumbens. *Proc. Natl. Acad. Sci. USA* **106,** 2915–2920.

Krause, J. E., Reiner, A. J., Advis, J. P., and McKelvy, J. F. (1984). *In vivo* biosynthesis of [35S]- and [3H]substance P in the striatum of the rat and their axonal transport to the substantia nigra. *J. Neurosci.* **4,** 775–785.

Kreitzer, A. C., and Malenka, R. C. (2007). Endocannabinoid-mediated rescue of striatal LTD and motor deficits in Parkinson's disease models. *Nature* **445,** 643–647.

Kreitzer, A. C., and Malenka, R. C. (2008). Striatal plasticity and basal ganglia circuit function. *Neuron* **60,** 543–554.

Kuhlman, S. J., and Huang, Z. J. (2008). High-resolution labeling and functional manipulation of specific neuron types in mouse brain by Cre-activated viral gene expression. *PLoS ONE* **3,** e2005.

Kumar, A., Choi, K. H., Renthal, W., Tsankova, N. M., Theobald, D. E., Truong, H. T., Russo, S. J., Laplant, Q., Sasaki, T. S., Whistler, K. N., Neve, R. L., Self, D. W., *et al.* (2005). Chromatin remodeling is a key mechanism underlying cocaine-induced plasticity in striatum. *Neuron* **48,** 303–314.

Lachowicz, J. E., and Sibley, D. R. (1997). Molecular characteristics of mammalian dopamine receptors. *Pharmacol. Toxicol.* **81,** 105–113.

Lee, K. W., Kim, Y., Kim, A. M., Helmin, K., Nairn, A. C., and Greengard, P. (2006). Cocaine-induced dendritic spine formation in D1 and D2 dopamine receptor-containing medium spiny neurons in nucleus accumbens. *Proc. Natl. Acad. Sci. USA* **103,** 3399–3404.

Lemberger, T., Parlato, R., Dassesse, D., Westphal, M., Casanova, E., Turiault, M., Tronche, F., Schiffmann, S. N., and Schutz, G. (2007). Expression of Cre recombinase in dopaminoceptive neurons. *BMC Neurosci.* **8,** 4.

Le Moine, C., Normand, E., and Bloch, B. (1991). Phenotypical characterization of the rat striatal neurons expressing the D1 dopamine receptor gene. *Proc. Natl. Acad. Sci. USA* **88,** 4205–4209.

Le Moine, C., Normand, E., Guitteny, A. F., Fouque, B., Teoule, R., and Bloch, B. (1990). Dopamine receptor gene expression by enkephalin neurons in rat forebrain. *Proc. Natl. Acad. Sci. USA* **87,** 230–234.

Lett, B. T., Grant, V. L., Byrne, M. J., and Koh, M. T. (2000). Pairings of a distinctive chamber with the aftereffect of wheel running produce conditioned place preference. *Appetite* **34,** 87–94.

Levine, A. A., Guan, Z., Barco, A., Xu, S., Kandel, E. R., and Schwartz, J. H. (2005). CREB-binding protein controls response to cocaine by acetylating histones at the fosB promoter in the mouse striatum. *Proc. Natl. Acad. Sci. USA* **102,** 19186–19191.

Lin, H., and Grosschedl, R. (1995). Failure of B-cell differentiation in mice lacking the transcription factor EBF. *Nature* **376,** 263–267.

Lobo, M. K., Cui, Y., Ostlund, S. B., Balleine, B. W., and Yang, X. W. (2007). Genetic control of instrumental conditioning by striatopallidal neuron-specific S1P receptor Gpr6. *Nat. Neurosci.* **10,** 1395–1397.

Lobo, M. K., Karsten, S. L., Gray, M., Geschwind, D. H., and Yang, X. W. (2006). FACS-array profiling of striatal projection neuron subtypes in juvenile and adult mouse brains. *Nat. Neurosci.* **9,** 443–452.

Lobo, M. K., Yeh, C., and Yang, X. W. (2008). Pivotal role of early B-cell factor 1 in development of striatonigral medium spiny neurons in the matrix compartment. *J. Neurosci. Res.* **86,** 2134–2146.

Lu, X., Shen, Y., Dong, H.-W., Fan, G., and Yang, X. W. (2008). Genetic analyses of a striatonigral neuron enriched orphan nucleotide sugar transporter Slc35D3. *Soc. Neurosci.* **2008,** 578.20.

Marsh, E. D., Minarcik, J., Campbell, K., Brooks-Kayal, A. R., and Golden, J. A. (2008). FACS-array gene expression analysis during early development of mouse telencephalic interneurons. *Dev. Neurobiol.* **68,** 434–445.

Mason, H. A., Rakowiecki, S. M., Raftopoulou, M., Nery, S., Huang, Y., Gridley, T., and Fishell, G. (2005). Notch signaling coordinates the patterning of striatal compartments. *Development* **132,** 4247–4258.

Matamales, M., Bertran-Gonzalez, J., Salomon, L., Degos, B., Deniau, J. M., Valjent, E., Herve, D., and Girault, J. A. (2009). Striatal medium-sized spiny neurons: Identification by nuclear staining and study of neuronal subpopulations in BAC transgenic mice. *PLoS ONE* **4,** e4770.

McClung, C. A., Ulery, P. G., Perrotti, L. I., Zachariou, V., Berton, O., and Nestler, E. J. (2004). DeltaFosB: A molecular switch for long-term adaptation in the brain. *Brain Res. Mol. Brain Res.* **132,** 146–154.

McLean, S., Bannon, M. J., Zamir, N., and Pert, C. B. (1985). Comparison of the substance P- and dynorphin-containing projections to the substantia nigra: A radioimmunocytochemical and biochemical study. *Brain Res.* **361,** 185–192.

Menalled, L., Zanjani, H., MacKenzie, L., Koppel, A., Carpenter, E., Zeitlin, S., and Chesselet, M. F. (2000). Decrease in striatal enkephalin mRNA in mouse models of Huntington's disease. *Exp. Neurol.* **162,** 328–342.

Mermelstein, P. G., Foehring, R. C., Tkatch, T., Song, W. J., Baranauskas, G., and Surmeier, D. J. (1999). Properties of Q-type calcium channels in neostriatal and cortical neurons are correlated with beta subunit expression. *J. Neurosci.* **19,** 7268–7277.

Mermelstein, P. G., Song, W. J., Tkatch, T., Yan, Z., and Surmeier, D. J. (1998). Inwardly rectifying potassium (IRK) currents are correlated with IRK subunit expression in rat nucleus accumbens medium spiny neurons. *J. Neurosci.* **18,** 6650–6661.

Missale, C., Nash, S. R., Robinson, S. W., Jaber, M., and Caron, M. G. (1998). Dopamine receptors: From structure to function. *Physiol. Rev.* **78,** 189–225.

Mitchell, I. J., Cooper, A. J., and Griffiths, M. R. (1999). The selective vulnerability of striatopallidal neurons. *Prog. Neurobiol.* **59,** 691–719.

Moratalla, R., Vallejo, M., Elibol, B., and Graybiel, A. M. (1996). D1-class dopamine receptors influence cocaine-induced persistent expression of Fos-related proteins in striatum. *Neuroreport* **8,** 1–5.

Mroz, E. A., Brownstein, M. J., and Leeman, S. E. (1977). Evidence for substance P in the striato-nigral tract. *Brain Res.* **125,** 305–311.

Nakano, K. (2000). Neural circuits and topographic organization of the basal ganglia and related regions. *Brain Dev.* **22**(Suppl. 1), S5–16.

Nakano, K., Kayahara, T., Tsutsumi, T., and Ushiro, H. (2000). Neural circuits and functional organization of the striatum. *J. Neurol.* **247**(Suppl. 5), V1–15.

Nelson, S. B., Hempel, C., and Sugino, K. (2006). Probing the transcriptome of neuronal cell types. *Curr. Opin. Neurobiol.* **16,** 571–576.

Nestler, E. J. (2008). Review. Transcriptional mechanisms of addiction: Role of DeltaFosB. *Philos. Trans. R. Soc. Lond. B. Biol. Sci.* **363,** 3245–3255.

Nestler, E. J., Barrot, M., and Self, D. W. (2001). DeltaFosB: A sustained molecular switch for addiction. *Proc. Natl. Acad. Sci. USA* **98,** 11042–11046.

Nye, H. E., Hope, B. T., Kelz, M. B., Iadarola, M., and Nestler, E. J. (1995). Pharmacological studies of the regulation of chronic FOS-related antigen induction by cocaine in the striatum and nucleus accumbens. *J. Pharmacol. Exp. Ther.* **275,** 1671–1680.

O'Neill, L. P., VerMilyea, M. D., and Turner, B. M. (2006). Epigenetic characterization of the early embryo with a chromatin immunoprecipitation protocol applicable to small cell populations. *Nat. Genet.* **38,** 835–841.

Packard, M. G., and Knowlton, B. J. (2002). Learning and memory functions of the Basal Ganglia. *Annu. Rev. Neurosci.* **25,** 563–593.

Padmanabhan, S., Myers, A. G., and Prasad, B. M. (2009). Constitutively active GPR6 is located in the intracellular compartments. *FEBS. Lett.* **583,** 107–112.

Pecina, S., Smith, K. S., and Berridge, K. C. (2006). Hedonic hot spots in the brain. *Neuroscientist* **12,** 500–511.

Perrotti, L. I., Hadeishi, Y., Ulery, P. G., Barrot, M., Monteggia, L., Duman, R. S., and Nestler, E. J. (2004). Induction of deltaFosB in reward-related brain structures after chronic stress. *J. Neurosci.* **24,** 10594–10602.

Pulipparacharuvil, S., Renthal, W., Hale, C. F., Taniguchi, M., Xiao, G., Kumar, A., Russo, S. J., Sikder, D., Dewey, C. M., Davis, M. M., Greengard, P., Nairn, A. C., *et al.* (2008). Cocaine regulates MEF2 to control synaptic and behavioral plasticity. *Neuron* **59,** 621–633.

Renthal, W., Carle, T. L., Maze, I., Covington, H. E. 3rd., Truong, H. T., Alibhai, I., Kumar, A., Montgomery, R. L., Olson, E. N., and Nestler, E. J. (2008). Delta FosB mediates epigenetic desensitization of the c-fos gene after chronic amphetamine exposure. *J. Neurosci.* **28,** 7344–7349.

Renthal, W., Kumar, A., Xiao, G., Wilkinson, M., Covington, H. E. 3rd., Maze, I., Sikder, D., Robison, A. J., LaPlant, Q., Dietz, D. M., Russo, S. J., Vialou, V., *et al.* (2009). Genome-wide analysis of chromatin regulation by cocaine reveals a role for sirtuins. *Neuron* **62,** 335–348.

Renthal, W., Maze, I., Krishnan, V., Covington, H. E. 3rd., Xiao, G., Kumar, A., Russo, S. J., Graham, A., Tsankova, N., Kippin, T. E., Kerstetter, K. A., Neve, R. L., *et al.* (2007). Histone deacetylase 5 epigenetically controls behavioral adaptations to chronic emotional stimuli. *Neuron* **56,** 517–529.

Richardson, P. J., Kase, H., and Jenner, P. G. (1997). Adenosine A2A receptor antagonists as new agents for the treatment of Parkinson's disease. *Trends Pharmacol. Sci.* **18,** 338–344.

Russo, S. J., *et al.* (2009). Nuclear factor kappa B signaling regulates neuronal morphology and cocaine reward. *J. Neurosci.* **29,** 3529–3537.

Saka, E., and Graybiel, A. M. (2003). Pathophysiology of Tourette's syndrome: Striatal pathways revisited. *Brain Dev.* **25**(Suppl. 1), S15–S19.

Sano, H., Yasoshima, Y., Matsushita, N., Kaneko, T., Kohno, K., Pastan, I., and Kobayashi, K. (2003). Conditional ablation of striatal neuronal types containing dopamine D2 receptor disturbs coordination of basal ganglia function. *J. Neurosci.* **23,** 9078–9088.

Santiago, M. P., and Potter, L. T. (2001). Biotinylated m4-toxin demonstrates more M4 muscarinic receptor protein on direct than indirect striatal projection neurons. *Brain Res.* **894,** 12–20.

Santini, E., Valjent, E., Usiello, A., Carta, M., Borgkvist, A., Girault, J. A., Herve, D., Greengard, P., and Fisone, G. (2007). Critical involvement of cAMP/DARPP-32 and extracellular signal-regulated protein kinase signaling in L-DOPA-induced dyskinesia. *J. Neurosci.* **27,** 6995–7005.

Santini, E., Alcacer, C., Cacciatore, S., Heiman, M., Herve, D., Greengard, P., Girault, J. A., Valjent, E., and Fisone, G. (2009). L-DOPA activates ERK signaling and phosphorylates histone H3 in the striatonigral medium spiny neurons of hemiparkinsonian mice. *J. Neurochem.* **108,** 621–633.

Schiffmann, S. N., and Vanderhaeghen, J. J. (1993). Adenosine A2 receptors regulate the gene expression of striatopallidal and striatonigral neurons. *J. Neurosci.* **13,** 1080–1087.

Schiffmann, S. N., Libert, F., Vassart, G., Dumont, J. E., and Vanderhaeghen, J. J. (1990). A cloned G protein-coupled protein with a distribution restricted to striatal medium-sized neurons. Possible relationship with D1 dopamine receptor. *Brain Res.* **519,** 333–337.

Schiffmann, S. N., Jacobs, O., and Vanderhaeghen, J. J. (1991a). Striatal restricted adenosine A2 receptor (RDC8) is expressed by enkephalin but not by substance P neurons: An *in situ* hybridization histochemistry study. *J. Neurochem.* **57,** 1062–1067.

Schiffmann, S. N., Libert, F., Vassart, G., and Vanderhaeghen, J. J. (1991b). Distribution of adenosine A2 receptor mRNA in the human brain. *Neurosci. Lett.* **130,** 177–181.

Schultz, W. (2002). Getting formal with dopamine and reward. *Neuron* **36,** 241–263.

Schuster, S., Nadjar, A., Guo, J. T., Li, Q., Ittrich, C., Hengerer, B., and Bezard, E. (2008). The 3-hydroxy-3-methylglutaryl-CoA reductase inhibitor lovastatin reduces severity of L-DOPA-induced abnormal involuntary movements in experimental Parkinson's disease. *J. Neurosci.* **28,** 4311–4316.

Schwarzschild, M. A., Agnati, L., Fuxe, K., Chen, J. F., and Morelli, M. (2006). Targeting adenosine A2A receptors in Parkinson's disease. *Trends Neurosci.* **29,** 647–654.

Selva, E. M., Hong, K., Baeg, G. H., Beverley, S. M., Turco, S. J., Perrimon, N., and Hacker, U. (2001). Dual role of the fringe connection gene in both heparan sulphate and fringe-dependent signalling events. *Nat. Cell Biol.* **3,** 809–815.

Shaner, N. C., Campbell, R. E., Steinbach, P. A., Giepmans, B. N., Palmer, A. E., and Tsien, R. Y. (2004). Improved monomeric red, orange and yellow fluorescent proteins derived from *Discosoma* sp. red fluorescent protein. *Nat. Biotechnol.* **22,** 1567–1572.

Shen, W., Flajolet, M., Greengard, P., and Surmeier, D. J. (2008). Dichotomous dopaminergic control of striatal synaptic plasticity. *Science* **321,** 848–851.

Shen, W., Tian, X., Day, M., Ulrich, S., Tkatch, T., Nathanson, N. M., and Surmeier, D. J. (2007). Cholinergic modulation of Kir2 channels selectively elevates dendritic excitability in striatopallidal neurons. *Nat. Neurosci.* **10,** 1458–1466.

Shuen, J. A., Chen, M., Gloss, B., and Calakos, N. (2008). Drd1a-tdTomato BAC transgenic mice for simultaneous visualization of medium spiny neurons in the direct and indirect pathways of the basal ganglia. *J. Neurosci.* **28,** 2681–2685.

Stefani, A., Chen, Q., Flores-Hernandez, J., Jiao, Y., Reiner, A., and Surmeier, D. J. (1998). Physiological and molecular properties of AMPA/Kainate receptors expressed by striatal medium spiny neurons. *Dev. Neurosci.* **20,** 242–252.

Sugino, K., Hempel, C. M., Miller, M. N., Hattox, A. M., Shapiro, P., Wu, C., Huang, Z. J., and Nelson, S. B. (2006). Molecular taxonomy of major neuronal classes in the adult mouse forebrain. *Nat. Neurosci.* **9,** 99–107.

Sun, Z., Del Mar, N., Meade, C., Goldowitz, D., and Reiner, A. (2002). Differential changes in striatal projection neurons in R6/2 transgenic mice for Huntington's disease. *Neurobiol. Dis.* **11,** 369–385.

Surmeier, D. J., Song, W. J., and Yan, Z. (1996). Coordinated expression of dopamine receptors in neostriatal medium spiny neurons. *J. Neurosci.* **16,** 6579–6591.

Surmeier, D. J., Ding, J., Day, M., Wang, Z., and Shen, W. (2007). D1 and D2 dopamine-receptor modulation of striatal glutamatergic signaling in striatal medium spiny neurons. *Trends Neurosci.* **30,** 228–235.

Svenningsson, P., Le Moine, C., Fisone, G., and Fredholm, B. B. (1999). Distribution, biochemistry and function of striatal adenosine A2A receptors. *Prog. Neurobiol.* **59,** 355–396.

Svenningsson, P., Nairn, A. C., and Greengard, P. (2005). DARPP-32 mediates the actions of multiple drugs of abuse. *AAPS. J.* **7,** E353–E360.

Taverna, S., Ilijic, E., and Surmeier, D. J. (2008). Recurrent collateral connections of striatal medium spiny neurons are disrupted in models of Parkinson's disease. *J. Neurosci.* **28,** 5504–5512.

Tietjen, I., Rihel, J. M., Cao, Y., Koentges, G., Zakhary, L., and Dulac, C. (2003). Single-cell transcriptional analysis of neuronal progenitors. *Neuron* **38,** 161–175.

Tietjen, I., Rihel, J., and Dulac, C. G. (2005). Single-cell transcriptional profiles and spatial patterning of the mammalian olfactory epithelium. *Int. J. Dev. Biol.* **49,** 201–207.

Toresson, H., and Campbell, K. (2001). A role for Gsh1 in the developing striatum and olfactory bulb of Gsh2 mutant mice. *Development* **128,** 4769–4780.

Tsai, H. C., Zhang, F., Adamantidis, A., Stuber, G. D., Bonci, A., de Lecea, L., and Deisseroth, K. (2009). Phasic firing in dopaminergic neurons is sufficient for behavioral conditioning. *Science* **324,** 1080–1084.

Uhlenbrock, K., Gassenhuber, H., and Kostenis, E. (2002). Sphingosine 1-phosphate is a ligand of the human gpr3, gpr6 and gpr12 family of constitutively active G protein-coupled receptors. *Cell. Signal.* **14,** 941–953.

Valjent, E., Corvol, J. C., Pages, C., Besson, M. J., Maldonado, R., and Caboche, J. (2000). Involvement of the extracellular signal-regulated kinase cascade for cocaine-rewarding properties. *J. Neurosci.* **20,** 8701–8709.

van der Kooy, D., and Fishell, G. (1987). Neuronal birthdate underlies the development of striatal compartments. *Brain Res.* **401,** 155–161.

Vilaro, M. T., Wiederhold, K. H., Palacios, J. M., and Mengod, G. (1991). Muscarinic cholinergic receptors in the rat caudate-putamen and olfactory tubercle belong predominantly to the m4 class: *In situ* hybridization and receptor autoradiography evidence. *Neuroscience* **40,** 159–167.

Wallace, D. L., Vialou, V., Rios, L., Carle-Florence, T. L., Chakravarty, S., Kumar, A., Graham, D. L., Green, T. A., Kirk, A., Iniguez, S. D., Perrotti, L. I., Barrot, M., *et al.* (2008). The influence of DeltaFosB in the nucleus accumbens on natural reward-related behavior. *J. Neurosci.* **28,** 10272–10277.

Wang, H. B., Laverghetta, A. V., Foehring, R., Deng, Y. P., Sun, Z., Yamamoto, K., Lei, W. L., Jiao, Y., and Reiner, A. (2006a). Single-cell RT-PCR, *in situ* hybridization histochemical, and immunohistochemical studies of substance P and enkephalin co-occurrence in striatal projection neurons in rats. *J. Chem. Neuroanat.* **31,** 178–199.

Wang, Z., Kai, L., Day, M., Ronesi, J., Yin, H. H., Ding, J., Tkatch, T., Lovinger, D. M., and Surmeier, D. J. (2006b). Dopaminergic control of corticostriatal long-term synaptic depression in medium spiny neurons is mediated by cholinergic interneurons. *Neuron* **50,** 443–452.

Warming, S., Liu, P., Suzuki, T., Akagi, K., Lindtner, S., Pavlakis, G. N., Jenkins, N. A., and Copeland, N. G. (2003). Evi3, a common retroviral integration site in murine B-cell lymphoma, encodes an EBFAZ-related Kruppel-like zinc finger protein. *Blood* **101,** 1934–1940.

Weiner, D. M., Levey, A. I., and Brann, M. R. (1990). Expression of muscarinic acetylcholine and dopamine receptor mRNAs in rat basal ganglia. *Proc. Natl. Acad. Sci. USA* **87,** 7050–7054.

Werme, M., Messer, C., Olson, L., Gilden, L., Thoren, P., Nestler, E. J., and Brene, S. (2002). Delta FosB regulates wheel running. *J. Neurosci.* **22,** 8133–8138.

Wilkinson, M. B., Xiao, G., Kumar, A., LaPlant, Q., Renthal, W., Sikder, D., Kodadek, T. J., and Nestler, E. J. (2009). Imipramine treatment and resiliency exhibit similar chromatin regulation in the mouse nucleus accumbens in depression models. *J. Neurosci.* **29,** 7820–7832.

Wojcik, W. J., and Neff, N. H. (1983a). Location of adenosine release and adenosine A2 receptors to rat striatal neurons. *Life Sci.* **33,** 755–763.

Wojcik, W. J., and Neff, N. H. (1983b). Differential location of adenosine A1 and A2 receptors in striatum. *Neurosci. Lett.* **41,** 55–60.

Wojcik, W. J., and Neff, N. H. (1983c). Adenosine A1 receptors are associated with cerebellar granule cells. *J. Neurochem.* **41,** 759–763.

Yan, Z., Flores-Hernandez, J., and Surmeier, D. J. (2001). Coordinated expression of muscarinic receptor messenger RNAs in striatal medium spiny neurons. *Neuroscience* **103,** 1017–1024.

Yang, X. W., Model, P., and Heintz, N. (1997). Homologous recombination based modification in Escherichia coli and germline transmission in transgenic mice of a bacterial artificial chromosome. *Nat. Biotechnol.* **15,** 859–865.

Yin, H. H., and Knowlton, B. J. (2006). The role of the basal ganglia in habit formation. *Nat. Rev. Neurosci.* **7,** 464–476.

Yun, K., Garel, S., Fischman, S., and Rubenstein, J. L. (2003). Patterning of the lateral ganglionic eminence by the Gsh1 and Gsh2 homeobox genes regulates striatal and olfactory bulb histogenesis and the growth of axons through the basal ganglia. *J. Comp. Neurol.* **461,** 151–165.

Zachariou, V., Bolanos, C. A., Selley, D. E., Theobald, D., Cassidy, M. P., Kelz, M. B., Shaw-Lutchman, T., Berton, O., Sim-Selley, L. J., Dileone, R. J., Kumar, A., and Nestler, E. J. (2006). An essential role for DeltaFosB in the nucleus accumbens in morphine action. *Nat. Neurosci.* **9,** 205–211.

BAC TO DEGENERATION: BACTERIAL ARTIFICIAL CHROMOSOME (BAC)-MEDIATED TRANSGENESIS FOR MODELING BASAL GANGLIA NEURODEGENERATIVE DISORDERS

Xiao-Hong Lu

Department of Psychiatry & Biobehavioral Sciences, Center for Neurobehavioral Genetics, Semel Institute for Neuroscience & Human Behavior, Brain Research Institute, David Geffen School of Medicine at UCLA, Los Angeles, California 90095, USA

Basal ganglia neurodegenerative disorders, such as Parkinson's disease (PD) and Huntington's disease (HD), are characterized by not only spectrum of motor deficits, ranging form hypokinesia to hyperkinesia, but also emotional, cognitive, and psychiatric manifestations. The symptoms and pathogenic mechanism of these disorders should be viewed as dysfunctions of specific cortico-subcortical neurocircuits. Transgenic approaches using large genomic inserts, such as bacterial artificial chromosome (BAC)-mediated transgeneis, due to its capacity to propagate large-size genomic DNA and faithful production of endogenous-like gene expression pattern/lever, have provided an ideal basis for the generation of transgenic mice as model for basal ganglia neurodegenerative disorders, as well as the functional and structural analysis of neurocircuits. In this chapter, the basic concepts and practical approaches about application of BAC transgenic system are introduced. Existent major BAC transgenic mouse

INTERNATIONAL REVIEW OF NEUROBIOLOGY, VOL. 89
DOI: 10.1016/S0074-7742(09)89002-8

0074-7742/09 $35.00

models for PD and HD are evaluated according to their construct, face, and predicative validity. Finally, considerations, possible solutions, and future perspectives of using BAC transgenic approach to study basal ganglia neurodegenerative disorders are discussed.

I. Introduction

Basal ganglia is a large collections of subcortical nuclei interconnected with the cerebral cortex, thalamus, and brainstem. Physiological function of basal ganglia has not only related to movement control but also has been implicated in cognitive function, emotion, and learning (Graybiel, 2000). A lot of human basal ganglia-related neurodegenerative disorders, such as Parkinson's disease (PD) and Huntington's disease (HD), are characterized by not only spectrum of motor deficits, ranging form hypokinesia to hyperkinesia, but also emotional, cognitive, and psychiatric manifestations (Thomas and Beal, 2007; Vonsattel and DiFiglia, 1998). The symptoms and pathogenic mechanism of these disorders should be viewed as the dysfunction of specific cortico-subcortical neurocircuits. These circuits comprise reentrant pathways that originate in various neocortical domains and pass through the basal ganglia before returning, by way of the thalamus, to select portions of the frontal lobe (Albin *et al.*, 1989). The major revisions of the views about anatomy and function (dysfunction) of basal ganglia in the last decade demand new perspectives and methodology to dissect the functions of the circuitries and to understand the role of these structures in various neuropsychiatric disorders.

In the last decade, bacterial artificial chromosomes (BACs) transgenic technology, due to its capacity to propagate large size genomic DNA and faithful production of endogenous-like gene expression pattern and lever, has provided an ideal basis for the generation of transgenic mice as model for basal ganglia neurodegenerative disorders as well as the functional and structural analysis of neurocircuitry (Heintz, 2001; Yang and Gong, 2005). Furthermore, the advent of large-scale transgenic projects based on BAC transgenesis, such as the Gene Expression Nervous System Atlas (GENSAT) BAC transgenic project, further increases the usefulness of BAC transgenesis to address above questions and leads to a boom of generation of BAC transgenic mice (Heintz, 2004). Many laboratories have successfully applied the BAC-mediated transgenic technology to generate animal models of PD and HD with robust phenotypes and are used in preclinical therapeutics study. In this chapter, the basic concepts, practical approaches about application of BAC transgenic system, and current major BAC mouse models in studying PD and HD are reviewed, along with a discussion about limitations and possible solutions to BAC transgenic approach.

II. Concept and Advantages of the BAC Transgenesis Approach

In 1976, Jaenisch *et al.* generated the first transgenic mice by introducing viral genomic DNA into the germline of mice *in vivo* (Jaenisch, 1977). Over the last decade, transgenic mouse models have become a very useful genetic tool to overexpress an exogenous transgene, in order to assess the phenotypic consequences *in vivo*. During this time, there has been a growing expectation that transgenes resemble the *in vivo* state as much as possible.

However, the traditional transgenic studies, using genomic DNA elements less than 20 kilobases (kb), often results in variable, ectopic, or even failure of transgene expression due to the positional effects, because the same transgene is integrated at different genomic loci. The positional effects are caused by limited size of the genomic DNA fragments that lack critical transcriptional regulatory elements. Positional effects can be overcome by using very large genomic DNA constructs (>100 kb), which likely contain most of the regulatory elements for a gene of interest (Giraldo and Montoliu, 2001).

In the early 1990s, several groups generated transgenic mice with yeast artificial chromosomes (YACs) and BACs (Jakobovits *et al.*, 1993; Yang *et al.*, 1997). YACs contain an average of 500–600 kb genomic DNA, and are propagated in yeast as linear artificial chromosomes. YAC transgenesis provide the initial proof-of-principle that large genomic inserts can confer integration-site-independent and dosage-dependent transgene expression (Forget, 1993; Peterson *et al.*, 1997). BACs are a circular, bacterial F-element-based plasmid containing an average of about 180–200 kb of genomic DNA. The BAC system has several advantages over the YAC system, including clonal stability, low chimerism, and the ability to obtain a large quantity of high quality DNA (Yang and Gong, 2005). For the majority of the genes in a mammalian genome (particularly for those that are <100 kb), an appropriately designed BAC transgene has a very high probability to confer accurate transgene expression *in vivo*. Numerous studies demonstrate that BAC transgenes could direct gene expression at physiological levels with the same developmental timing and expression patterns as endogenous genes in transgenic animal models (Yang, 2008). Furthermore, a lot of large-scale projects have emerged based on BAC transgenesis, for example, GENSAT project aims at mapping all genes expressed in the central nervous system; and VelociGene, Regeneron aims at characterizing potential targets for drug development. These developments lead to a boom in the number of BAC transgenic mice generated in the last decade.

III. Generation of the BAC Transgenic Mice via Homologous Recombination

The first method developed to engineer BACs is the RecA-based method (Yang *et al.*, 1997). The original RecA-based method, as well as its subsequent improved version (Gong *et al.*, 2002), utilizes a shuttle vector that temporarily introduces the *Escherichia coli* RecA gene into the BAC-host bacteria to restore its recombination competence. The detailed methods for engineering the BAC using the RecA methods could be found by Gong and Yang (2005).

The second widely used method for BAC engineering is the recombineering method, also referred to as λ-mediated recombination. The BAC is modified by homologous recombination in strains of *E. coli* which express phage-derived protein pairs, either RecE/RecT from the Rac prophage (Zhang *et al.*, 1998), or Redα/Redβ from λ phage (Copeland *et al.*, 2001; Lee *et al.*, 2001). Both the RecA-based and the recombineering methods are highly efficient and reliable methods to introduce mutations or marker genes into a BAC. These methods have both been applied successfully in large-scale BAC engineering projects.

IV. Technique Considerations and Possible Solutions

A. Integration Sites and Copy Number

One of the major considerations of BAC transgenic technology is that the integration BAC transgene is random and uncontrolled. This generates three problems. First, random integration of BAC transgene may cause insertional mutagenesis to inactivate the endogenous gene at the site of integration. Second, the BAC transgene may simply insert into a silent locus, and therefore, the expression of transgene may be shut off. Third, BAC transgene integration usually happens in a singular event with multiple copies inserted in tandem. The high tandem copy number may lead to gene silencing (Ristevski, 2005). Several measures could be applied to reduce above possibilities. For example, at least two transgenic lines should be analyzed to differentiate transgene specific effects from the positional effects. In addition, cytogenetics methods such as fluorescence *in situ* hybridization (FISH) could be used to screen BAC transgenic animals to identify the unexpected genetic events, such as transgene-induced chromosomal rearrangements (Abrahams *et al.*, 2003).

B. Size of the BAC

One limitation for the BAC transgenic approach applies to the relatively small number of mammalian genes that are a spanning above 200 kb in genomic size. For these genes, two or even three BACs may be needed to cover the whole genomic locus and its proper regulatory elements. Some strategies have been developed to overcome this limitation. BAC fusion methodology (Sopher and La Spada, 2006) can be easily used to recombine two overlapping BAC into one seamless larger BAC. The BAC fusion strategy is based on recombineering method. Briefly, first the BAC containing 5′ genomic locus of the gene of the interest is modified to introduce a kanamycin (Kan) selection marker at the 3′ end of this BAC, and the Kan gene is followed by a small homology box corresponding to about a 400 base pair region shared by the two BACs and located at the 3′ end of the other BAC containing the 3′ genomic locus of GOI. Next, 3′BAC is electroporated into DY380 bacteria which supports λ phage recombinase based "recombineering" (Lee *et al.*, 2001; Warming *et al.*, 2005). Afterward, restriction digest is performed to release the 5′BAC insert with Kan insert at 3′ end, and electroporate this BAC insert into the DY380 bacteria. Two recombination events will ensure the proper fusion of the two BACs to generate the seamless recombined BAC, which can be selected by growth on kanamycin and chloramphenicol.

C. BAC Purification and Microinjections

BAC transgenic mice are generated via direct microinjection of BAC DNA into the pronucleus of fertilized mouse eggs. Compared to plasmid DNA used for the traditional transgenic approach, preparing BAC DNA for microinjection is a more challenging task. There are several critical questions need to be determined before a microinjection, such as the purification of BAC DNA, method to examine the intact BAC DNA, linear or circular DNA, and concentration of injection. As for the method for the BAC DNA purification, the most successful method, in terms of consistency in generating transgenic founders, is the cesium chloride centrifugation method, which is described in detail elsewhere (Gong and Yang, 2005). Pulse-field gel electrophoresis (PFGE) analysis is suggested to examine the intactness of the purified BAC DNA (Gong and Yang, 2005). In addition, it was suggested that polyamine buffer has protective effects to BAC DNA and 0.5 ng/μl concentration is essential for the injection (Van Keuren *et al.*, 2009).

In the same study, no significant difference was observed between circular and linear BAC DNA (Van Keuren *et al.*, 2009).

D. Other Genes on the Same BAC

Due to the large size of the BAC transgene, sometimes it may contain other genes besides the gene of the interest, which may complicate the interpretation of the results. In this case, it is suggested that DNA construct should be designed to ensure deletion of the other genes on the BAC via homologous recombination. In addition, the expression pattern of possible other genes on the BAC could be examined via *in situ* hybridization.

In summary, the advantages of the BAC transgenic approach greatly outweighs its limitations, and it is for this reason that the BAC approach has become a method of choice to develop robust new animal models for human brain disorders and to develop novel therapeutic strategies.

V. BAC Transgenic Models of Basal Ganglia Neurodegenerative Disorders

Basal ganglia circuitry plays a key role in the regulation of voluntary movements as well as in behavioral control and cognitive functions (Albin *et al.*, 1989; Graybiel, 2000). In this chapter, we will focus more on the basal ganglia neurodegenerative disorders, such as PD and HD, which cause profound motor disturbances. In the last decade, transgenic mice using large genomic transgenes to model the HD and PD has made significant progress, and is anticipated that more promising therapies based on disease mechanisms revealed from these models will emerge.

The fundamental assumption of the approach using genetically modified animals to model neurodegenerative disorders is that the introduction of a genetic alteration identified from human genetics studies might recapitulate the full spectrum of the pathological and behavioral phenotypes via the same molecular pathogenic mechanisms as seen in human patients. These animal models can be used to study slow onset and the progression of behavioral and pathological phenotypes over a very long time period during the life span of the modeling animal, which cannot be accessed using acute/toxin induced disease models. This advantage makes the genetic animal modeling approach a highly useful tool to guide for the development of therapeutics that not only relive the symptoms but also slow or even stop the progression of the disease. It has been suggested that the significance/value of these models should be evaluated using three validity criteria (Fleming *et al.*, 2005; Yang, 2008). These criteria could be characterized by

(1) construct validity requires that the animal model shared the same etiology and underlying pathophysiological mechanisms as seen in human disease; (2) face validity examines the resemblance in symptoms of human patients; and (3) predictive validity requires that the animal model should have similar pharmacological responses as in the human patients. Since large genomic transgenic technologies such as YACs and BACs often contain the intact genomic locus and direct accurate, endogenous-like expression of the disease-causing proteins, the phenotypes from these models are often more accurate and robust compared to conventional transgenic models. For example, YAC transgenic models of HD, AD, and Spinalbulbarmuscular atrophy (SBMA) all exhibit robust disease-like features (Hodgson *et al.*, 1999; Pearson and Choi, 1993; Sopher *et al.*, 2004). Therefore, large genomic insert transgenic approach can provide advantages in modeling neurodegenerative disorders with higher construct, face and predicative validity than the traditional transgenic technology and has become the method of choice to model neurodegenerative disorder in basal ganglia. In the following section, a thorough review of the existent mouse models of PD and HD using large genomic inserts, with a focus on BAC transgenic mouse models will be discussed (Table I).

VI. BAC Transgenesis to Model Huntington's Disorder (HD)

HD is an autosomal dominant neurodegenerative disorder characterized pathologically by progressive degeneration in basal ganglia, specifically the striatum (caudate-putamen) (Imarisio *et al.*, 2008). Striatal medium spiny neurons (but not interneuron) are the predominant cellular types affected (Ross, 2004). Symptoms of HD include progressive motor impairments, cognitive decline, and psychiatric disturbances (Vonsattel and DiFiglia, 1998). Since the discovery of that HD is caused by a CAG repeat expansion (>35 repeats) in the *huntingtin* (*htt*) gene (1993) (Sharp *et al.*, 1995), multiple mouse genetic mouse models have been generated to facilitate the pathophysiology and drug development studies for this devastating brain disease (Menalled and Chesselet, 2002; Zoghbi and Orr, 2000).

R6/2 mouse line is the first genetic mouse model of HD, which expresses an exon 1 fragment of htt with a range of 148–153 repeats driven by IT15 prompter (Mangiarini *et al.*, 1996). The R6/2 mice line displays most robust phenotypes among the existent mouse HD models, such as wide occurrence of htt aggregates, behavioral deficits as early as 5 weeks, progressive atrophy in brain and striatal volume, significant degeneration of striatal neurons by 12 weeks, and death by 12–15 weeks (Carter *et al.*, 1999; Mangiarini *et al.*, 1996). Therefore, R6/2 mouse model has very good face validity. However, it has relatively poor construct

TABLE I

Major Transgenic Mouse Models of PD and HD Using Large Genomic Inserts

Disease		Construct validity	Face validity: Motor deficits	Face validity: Degeneration	Face validity: Cardinal pathology	Predictive validity	Reference
Parkinson's disease	Alpha-synuclein (PARK1,)	++++ PAC-mediated overexpression of WT human alpha-synuclein	N.D. N.D.	N.D.	Atypical	N.D.	Gispert *et al.* (2003)
	Parkin (PARK2)	+++ Overexpression of human mutant parkin-Q311X using dopamine transporter promoter in a BAC	+++++ Yes (progressive)	Yes (selective and progressive)	Atypical	N.D.	Lu *et al.* (2009)
	LRRK2	+++++ BAC-mediated overexpression of human Lrrk2 R1441G mutation	++++ Yes	Subtle defects	No	+++++ Motor deficits can be reversed by L-DOPA	Li *et al.* (2009)
Huntington's disease	Htt-128 CAGs	+++++ YAC-mediated overexpression of Htt-128 CAG repeats	++++ Yes (progressive)	Yes (selective and progressive)	Htt aggregates	N.D.	Hodgson *et al.* (1999)
	Htt-97 CAGs	+++++ BAC-mediated overexpression of floxed Htt-97 CAG repeats	++++ Yes (progressive)	Yes (selective and progressive)	Htt aggregates	N.D.	Gray *et al.* (2008)

validity because it overexpresses only a fragment of human htt, which limits its value in the pathological mechanism study of HD.

In a marked contrast with R6/2 mice, several lines of knock-in models of HD have also been generated to express mouse mutant htt (Menalled and Chesselet, 2002). As far as the construct validity, these mice are of the highest value. However, these mice in general have less robust behavorial and pathological phenotypes than R6/2 mice, therefore less face validity. The predicative validly of these mice is still need to be determined.

The mouse model of HD with the best balance of the construct and face validity are generated using large genomic inserts, such as YAC and BAC. Michael Hayden's group produced YAC transgenic mice expressing mutant htt. Mice from this line express the full-length human htt with a CAG repeat length of up to 128 (Hodgson *et al.*, 1999). By 12 months of age, these models exhibit progressive motor and pathological deficits and, in YAC128 mice, culminate in selective cortical and striatal atrophy and neurodegeneration (Slow *et al.*, 2003; Van Raamsdonk *et al.*, 2005).

Yang's group at UCLA has generated a BAC-mediated transgenic mouse model (BACHD) of HD expressing fl-mhtt with 97 glutamine repeats (Gray *et al.*, 2008). BACHD mice exhibit robust progressive motor, cognitive and psychiatric-like deficits, late-onset cortical and striatal atrophy, and dark neuron degeneration. Importantly, in BACHD mice, two loxP sites were positioned to flank the mhtt exon 1, so that the expression of fl-mhtt can be switched off by Cre recombinase. This design will facilitate the study of the critical pathophysiologic mechanisms of HD, such as noncell autonomous influence of cortex. Furthermore, Power analyses reveal the robustness of the behavioral and neuropathological phenotypes, suggesting BACHD as a good full-length-mhtt mouse model for preclinical studies guide the development of therapeutics. In general, BAC and YAC transgenic mouse models tend to have very high construct and face validity, which display progressive motor abnormalities as well as neurodegeration in the striatum. These symptoms all closely parallel human symptoms in HD, and provide evidence that these mice might accurately model many aspects of the disease.

VII. BAC Transgenesis to Model Parkinson's Disease (PD)

PD is the second most common form of neurodegenerative disorder with cardinal clinical features of resting tremor, rigidity, and bradykinesia (Thomas and Beal, 2007). The pathological hallmark of PD is a progressive and preferential loss of dopaminergic (DA) neurons in the substantial nigra pars compacta (SNc) (Rao *et al.*, 2006). At present, pharmacotherapy represented by levodopa or other DA agents as well as surgical approaches such as deep-brain stimulation can only provide symptomatic improvement of PD (Dauer and Przedborski, 2003; Rao *et al.*, 2006). Therapeutical strategies that can slow or stop the neurodegenerative process or even prevent the onset of the disease are currently not available. The development of such kind of therapeutics will have a deep impact on the treatment of PD. The endeavor is seriously hindered by the lack of genetic mammalian animal models to recapitulate the cardinal feature of PD.

However, the complex genetics of PD and the perplexing pathogenic role of the genetic mutation interacting with environmental factors make generation of rodent model for PD a more challenging task than that of modeling HD. Different from HD, PD is not a monogenic disorder. Although predominant form of PD cases are sporadic, in rare families, some genes (including alpha-synuclein, Parkin, Larrk2, pink 1, DJ-1, UCH-L1, etc.) have been identified to cause PD (Hardy *et al.*, 2006; Morris, 2005). For some of these genes, the inheritance pattern and the disease causative role are still controversial (Hardy *et al.*, 2006; Klein *et al.*, 2007). Therefore, the top priority for modeling PD right now is to fulfill the requirement of construct validity. This is a different strategy than that in HD field, in which major interests are put into the discovering therapeutics, and therefore, more emphasis of the face and predicative validity of the animal models. However, the same as in the HD field, BAC transgenic mouse models of PD have consistently demonstrated its advantages in generating mouse models with higher construct, face, and predicative validity than traditional transgenic approach.

A. BAC Transgenic Model of PD Based on Alpha-Synuclein (PARK1) Mutation

The first gene identified associated with PD is alpha-synuclein (SNCA) (Polymeropoulos *et al.*, 1997). SNCA has been identified as a major component of Lewy bodies and Lewy neurites in all PD patients (Spillantini *et al.*, 1997; Trojanowski and Lee, 1998). Two different missense mutations in SNCA, A53T, and A30P, have been associated with dominantly inherited, early-onset familiar form of PD (Polymeropoulos *et al.*, 1997). In addition, SNCA locus triplication has been shown to cause PD (Singleton *et al.*, 2003).

To model the dominant toxicity of SNCA, multiple lines of transgenic mouse models have been generated with traditional transgenic technologies, to overexpress the wild type (WT) or mutant SNCA using a variety of promoters. Some representative models are summarized as follows: in models generated with the Prion promoter, there are progressive motor deficits, fibrillar inclusions of SNCA, and neurodegeneration, but the neuropathology is more restricted to the motor neurons in the brain stem and the spinal cord and no degeneration of DA neurons (Fernagut and Chesselet, 2004; Giasson *et al.*, 2002). Several models with Thy-1 promoter did not show DA neuron pathology since this promoter did not express in the DA neurons (Kahle *et al.*, 2000; van der Putten *et al.*, 2000). Two mouse models using tyrosine hydroxylase promoter to target alpha-synuclein in the DA neurons showed some accumulation of SNCA in DA neuron, but without Lewy body-like inclusions and DA neuron loss (Matsuoka *et al.*, 2001; Richfield *et al.*, 2002). In a mouse model overexpressing WT SNCA using PDGF promoter, DA neuron terminals and reduced tyrosine hydroxylase expression were observed, however, without overt DA neuron loss (Masliah *et al.*, 2000). In summary, these transgenic mouse models of PD have limited construct validity due to ectopic and sometimes extremely high level of overexpression of mutant/WT alpha-synuclein. As for the phenotypes (face validity), these models recapitulate synucelinopathy, one of the characteristic features of PD, but with very mild DA neuron pathology and Parkinsonism-like phenotypes.

Some initial efforts have been made to generate SNCA model of PD using transgenic animal model with large genomic inserts. Nussbaum's group has generated a P1 artificial chromosome (PAC)-mediated transgenic mice expressing WT human SNCA (Gispert *et al.*, 2003). The homozygous PAC mice have about 20 copies of the transgenes. A very interesting observation with this mouse model is that even with the whooping high-copy number of the transgenes, there is no Lewy body like SNCA inclusion observed in these mice. No additional pathological and behavioral deficits were reported for this mouse model. The results derived from this model is in contrast to the observation in the previous SNCA transgenic mice with PDGF promoter (Masliah *et al.*, 2000), suggesting that a different expression pattern of SNCA may elicit different pathogenic mechanism and consequences *in vivo*.

Several groups are now generating SNCA transgenic mouse model using BAC transgenic approach. Wade–Martins's group in the UK (Anwar, 2008 SFN abstract) generated a BAC transgenic-mediated A30P mutation-based model for PD. Transgenic founder mice were generated on a pure C57BL6 background and backcrossed onto an alpha-synuclein null-background to create the *SNCA-A30P* + *Snca*$^{-/-}$ line. The characterizations of this model are still going on, with a focus on the role of mutant alpha-synuclein in regulating dopamine neurochemistry, and the kinetics of dopamine release and reuptake and the presynaptic regulation of vesicular trafficking mechanisms. In this interesting model, surprisingly, there is a highly physiologically relevant expression pattern of the SNCA A30P transgene, most notably in cell bodies in the SNC. This observation is consistent with the report that synucleinopathy is exacerbated in mice expressing A53T SNCA on an SNCA null background (Cabin *et al.*, 2005). One possible explanation is that SNCA is under tight expression–regulation in DA cells. The existence of endogenous SNCA may downregulate the transgene expression. Therefore, SNCA transgene expression will present in the correct pattern only when it is in a null background. Several laboratories are pursuing generating SNCA transgenic mouse model in the null background, Dr. Nussbaum's group is breeding different lines of BAC transgenic mice expressing the normal or mutated forms of human SNCA in the absence of the mouse's own SNCA gene (personal communication). The resulting mice will be followed for 2 years to see if those with a double-dose of mutated SNCA is developing characteristic phenotypes of PD.

Furthermore, according to personal communication, Chenjian Li *et al.*, from Weill Medical College of Cornell University, will use BAC technology to create a transgenic rat overexpressing WT and mutant SNCA. Rat models of PD will provide useful models to study genetic environmental interaction of alpha-syncuelin with several environmental toxins (such as rotenone), which only produce consistent DA neurodegeneration in rats. In general, the SNCA BAC transgenic rodent models have the best construct validity among the genetic models of PD, because of the less controversial role of SNCA in the pathogenesis of PD. Therefore, these models are important tools to study pathogenic mechanisms and examine different hypothesis, such as the progression of synucelinopathy (Brahhak hypothesis). However, the face and predicative validity of these models are yet to be determined.

B. BAC Transgenic Model of PD Based on Parkin (PARK2) Mutation

Parkin gene (PARK2) mutations appear to be the most frequent cause of familial PD (Bertoli-Avella *et al.*, 2004). Parkin mutations were initially identified to be associated with autosomal recessive juvenile parkinsonism. Current hypothesis suggests that pathogenic Parkin mutations may cause a loss-of-function phenotype, which impairs the removal of potentially toxic Parkin substrates, and such substrates accumulate and induce dysfunction and degeneration of the DA neurons (Hattori and Mizuno, 2004; Imai and Takahashi, 2004). Hypothesis-driven searches have led to the generation of several Parkin knockout animal models. There are at least five major Parkin-deficient mouse models (including spontaneous mutant quaking mice) have been reported (Goldberg *et al.*, 2003; Itier *et al.*, 2003; Palacino *et al.*, 2004; Perez and Palmiter, 2005). Parkin-deficient mice displayed mild dysfunction of the DA and noradrenergic systems and related behavioral changes. However, Parkin knockout mice do not have overt DA neuron degeneration. In short, these mice have very high construct validity, based on hypothetic recessive nature of the mutation, but significant less-face validity. One of the approaches to further characterize these parkin-deficient animals would be using BAC transgenesis to overexpress WT Parkin. Cross of WT parkin mice with knockout mice will help to further characterize the function and explore the possible pathogenic mechanism of Parkin protein.

Emerging evidence also points at the possibility that certain *parkin* mutants may exert dominant effects (either gain-of-function or dominant negative effects), which may contribute to PD pathogenesis. First, patients with *parkin* heterozygous mutations exhibit DA neuron dysfunction in imaging studies (PET, fMRI, and SPECT) (Hilker *et al.*, 2001; Khan *et al.*, 2005; Varrone *et al.*, 2004). Furthermore, these heterozygous mutations may confer higher risk (Clark *et al.*, 2006) and reduce the age-of-onset in a collection of PD families (Sun *et al.*, 2006). Rare *parkin* mutations can also be inherited in an autosomal dominant manner (Kobayashi *et al.*, 2003). Several tissue culture studies found that not all Parkin mutations diminish the enzymatic function of Parkin. Some mutations (such as Q311X truncation mutation) displayed dominant effects such as decreased solubility, altered subcellular localization to form inclusion body, increased substrate binding affinity, sensitize cells to mitochondria toxins/oxidative stress, and increased apoptosis (Cookson *et al.*, 2003; Henn *et al.*, 2005; Hyun *et al.*, 2002, 2005; Sriram *et al.*, 2005; von Coelln *et al.*, 2004; Wang *et al.* 2005a,b). This evidence leads to an important yet less explored idea that besides loss of Parkin function, some parkin mutations may have dominant pathogenic effects to DA neurons. Thus, the animal models that overexpress the Parkin mutations rather than simply deleting parkin gene locus may provide an alternative to analyze the pathogenic function Parkin mutation. Along this line of works, dominant effects of mutant Parkin were observed *in vivo* in *Caenorhabditis elegans* and in *Drosophila.* A truncated *C. elegans* parkin homolog couples alpha-synuclein aggregation to proteotoxic stress and leads to temperature sensitive lethality in *C. elegans* (Springer *et al.*, 2005). The overexpression of human Parkin mutants, R275W, T240R, and Q311X, in *Drosophila* produced phenotypes including progressive motor deficits and selective degeneration of DA neurons (Sang *et al.*, 2007; Wang *et al.*, 2007).

To explore the possibility that mutated Parkin could exert dominant effects on DA neurons in mammalian models, we overexpressed a C-terminal truncated (amino acid position 311) parkin protein driven by a dopamine transporter (DAT) BAC (Lu *et al.*, 2009). Human Q311X mutation is a nonsense mutation that replaces a Gln codon (CAG) with a stop codon (TAG) at amino acid position 311 in exon 8 (Hattori *et al.*, 1998). The mutation has been speculated to result in a premature termination of translation to produce a truncated parkin protein that lacks 155 amino acid including the second RING finger motif and IBR domain. Parkin-Q311X mice exhibit multiple late-onset and progressive hypokinetic motor deficits. Stereological analyses reveal that the mutant mice develop age-dependent DA neuron degeneration in substantia nigra accompanied by a significant loss of DA neuron terminals in the striatum. Neurochemical analyses reveal a significant reduction of the striatal dopamine level in mutant mice, which is significantly correlated with their hypokinetic motor deficits. Finally, mutant Parkin-Q311X mice, but not WT controls, exhibit age-dependent accumulation of proteinase K-resistant endogenous alpha-synuclein in substantia nigra and colocalized with 3-nitrotyrosine, a marker for oxidative protein damage. Therefore, our Parkin-Q311 BAC transgenic mouse models provide the first parkin-related genetic model that recapitulates the "hallmark feature" of PD and have the best face validity that recapitulates cardinal symptoms of PD. Furthermore, this mouse model has been highlighted by SciBX as an important model to guide the development of the therapeutics for PD. However, the construct validity of the mouse is limited due to the targeted expression of the transgene in DA neuron, which prevents the studying selective DA neuron specific toxicity of this particular mutation. And the further L-DOPA rescue of the behavioral phenotypes is needed to establish its predicative validity.

C. BAC Transgenic Model of PD Based on Leucine-Rich Repeat Kinase 2 (LRRK2) Mutations

Leucine-rich repeat kinase 2 also known as LRRK2 is a protein member of the leucine-rich repeat kinase family (Shen, 2004). Researchers have identified at least 20 LRRK2 mutations in families with late-onset PD (Thomas and Beal, 2007; Zimprich *et al.*, 2004). It is still unclear how LRRK2 mutations lead to PD.

Using stringent BAC transgenic mouse genetics approach (two transgenic lines and WT BAC control were generated to exclude the possibilities of various confounding factors), Li *et al.* generated the first BAC transgenic mouse model of LRRK2 mutation-related PD (Li *et al.*, 2009). This mouse model exhibits striking behavioral and pathological phenotypes that mimic symptoms seen in human PD patients. These phenotypes include age-dependent hypokinesia, L-DOPA responsiveness, tauopathy, nigra–striatal dopamine neuron pathology, and related neurochemical dysfunctions. With solid construct, face and predicative validity, LRRK2-R1441G BAC transgenic mouse model meets the three requirements of a "bona-fide" PD animal model. Furthermore, this rather exciting paper represents a big step forward in the establishment of the causative role Lrrk2 mutation in PD pathogenesis.

A very intriguing observation of this model is the robust hypokinetic motor deficits and relatively milder nigra and striatal neuropathology, which is in marked contrast with the traditional toxin model, but is consistent with the recent reports in chronic toxin model (Fernagut *et al.*, 2007) and some genetic mouse models, such as Tansey–Goldberg's parkin knockout mice (Frank-Cannon *et al.*, 2008) and our Parkin truncation model (Lu *et al.*, 2009). These studies highlight the role of compensatory function of nigra–striatal neuron in the manifestation PD behavioral and pathological symptoms.

Overall, the LRRK2-R1441G BAC transgenic mouse model provides a valuable platform to test various pathogenic hypothesis of PD, such as environmental influence, interaction with other PD-related genes, oxidative stress/mitochondria pathways, etc., and most importantly, to guide the development of therapeutics to slow or even stop the progress of the symptoms of PD.

VIII. Modeling Neurodegenerative Diseases via BAC Transgenesis, A Future Perspective

The completion of the sequencing of the human genome and those of other organisms have provided unparalleled opportunities to drug discovery. One of the major tasks of functional genomics is to translate genetics into clinically useful drugs against a major human disease. Transgenic animal models using large genomic inserts, such as BAC, provide a powerful tool to translate the genetic information in the form of pathogenic mutations to study the etiology, disease progression, pathology, and testing potential therapeutic strategies. Furthermore, BAC transgenesis could be applied together with other genetic tools, such as Cre recombinase-mediated inducible expression system to control the expression of transgene in a conditional, temporal, and spatial specific patterns. BAC transgenesis can also be used to deliver RNA interference in a special neurocircuits to elicit posttranscriptional gene silencing. Most importantly, for neurodegenerative disorders, such as PD and HD, although genetic causes have been identified for a long time, the field is yet to develop therapeutic approach to slow or even stop the progress of the disease. Transgenic mouse model of HD and PD using BAC transgenesis will provide a relevant and relatively rapid system for determining the efficacy of therapeutic agents and their mechanisms of action and facilitate the subsequent selection of appropriate agents for large-scale clinical trials.

References

Abrahams, B. S., Chong, A. C., Nisha, M., Milette, D., Brewster, D. A., Berry, M. L., Muratkhodjaev, F., Mai, S., Rajcan-Separovic, E., and Simpson, E. M. (2003). Metaphase FISHing of transgenic mice recommended: FISH and SKY define BAC-mediated balanced translocation. *Genesis* **36,** 134–141.

Albin, R. L., Young, A. B., and Penney, J. B. (1989). The functional anatomy of basal ganglia disorders. *Trends Neurosci.* **12,** 366–375.

(1993). A novel gene containing a trinucleotide repeat that is expanded and unstable on Huntington's disease chromosomes. The Huntington's Disease Collaborative Research Group. *Cell* **72,** 971–983.

Bertoli-Avella, A. M., Oostra, B. A., and Heutink, P. (2004). Chasing genes in Alzheimer's and Parkinson's disease. *Hum. Genet.* **114,** 413–438.

Cabin, D. E., Gispert-Sanchez, S., Murphy, D., Auburger, G., Myers, R. R., and Nussbaum, R. L. (2005). Exacerbated synucleinopathy in mice expressing A53T SNCA on a Snca null background. *Neurobiol. Aging* **26,** 25–35.

Carter, R. J., Lione, L. A., Humby, T., Mangiarini, L., Mahal, A., Bates, G. P., Dunnett, S. B., and Morton, A. J. (1999). Characterization of progressive motor deficits in mice transgenic for the human Huntington's disease mutation. *J. Neurosci.* **19,** 3248–3257.

Clark, L. N., Afridi, S., Karlins, E., Wang, Y., Mejia-Santana, H., Harris, J., Louis, E. D., Cote, L. J., Andrews, H., Fahn, S., Waters, C., Ford, B., *et al.* (2006). Case-control study of the parkin gene in early-onset Parkinson disease. *Arch. Neurol.* **63,** 548–552.

Cookson, M. R., Lockhart, P. J., McLendon, C., O'Farrell, C., Schlossmacher, M., and Farrer, M. J. (2003). RING finger 1 mutations in Parkin produce altered localization of the protein. *Hum. Mol. Genet.* **12,** 2957–2965.

Copeland, N. G., Jenkins, N. A., and Court, D. L. (2001). Recombineering: A powerful new tool for mouse functional genomics. *Nat. Rev. Genet.* **2,** 769–779.

Dauer, W., and Przedborski, S. (2003). Parkinson's disease: Mechanisms and models. *Neuron* **39,** 889–909.

Fernagut, P. O., and Chesselet, M. F. (2004). Alpha-synuclein and transgenic mouse models. *Neurobiol. Dis.* **17,** 123–130.

Fernagut, P. O., Hutson, C. B., Fleming, S. M., Tetreaut, N. A., Salcedo, J., Masliah, E., and Chesselet, M. F. (2007). Behavioral and histopathological consequences of paraquat intoxication in mice: Effects of alpha-synuclein over-expression. *Synapse* **61,** 991–1001.

Fleming, S. M., Fernagut, P. O., and Chesselet, M. F. (2005). Genetic mouse models of parkinsonism: Strengths and limitations. *NeuroRx* **2,** 495–503.

Forget, B. G. (1993). YAC transgenes: Bigger is probably better. *Proc. Natl. Acad. Sci. USA* **90,** 7909–7911.

Frank-Cannon, T. C., Tran, T., Ruhn, K. A., Martinez, T. N., Hong, J., Marvin, M., Hartley, M., Trevino, I., O'Brien, D. E., Casey, B., Goldberg, M. S., and Tansey, M. G. (2008). Parkin deficiency increases vulnerability to inflammation-related nigral degeneration. *J. Neurosci.* **28,** 10825–10834.

Giasson, B. I., Duda, J. E., Quinn, S. M., Zhang, B., Trojanowski, J. Q., and Lee, V. M. (2002). Neuronal alpha-synucleinopathy with severe movement disorder in mice expressing A53T human alpha-synuclein. *Neuron* **34,** 521–533.

Giraldo, P., and Montoliu, L. (2001). Size matters: Use of YACs, BACs and PACs in transgenic animals. *Transgenic Res.* **10,** 83–103.

Gispert, S., Del Turco, D., Garrett, L., Chen, A., Bernard, D. J., Hamm-Clement, J., Korf, H. W., Deller, T., Braak, H., Auburger, G., and Nussbaum, R. L. (2003). Transgenic mice expressing mutant A53T human alpha-synuclein show neuronal dysfunction in the absence of aggregate formation. *Mol. Cell Neurosci.* **24,** 419–429.

Goldberg, M. S., Fleming, S. M., Palacino, J. J., Cepeda, C., Lam, H. A., Bhatnagar, A., Meloni, E. G., Wu, N., Ackerson, L. C., Klapstein, G. J., Gajendiran, M., Roth, B. L., *et al.* (2003). Parkin-deficient mice exhibit nigrostriatal deficits but not loss of dopaminergic neurons. *J. Biol. Chem.* **278,** 43628–43635.

Gong, S., and Yang, X. W. (2005). Modification of bacterial artificial chromosomes (BACs) and preparation of Intact BAC DNA for generation of transgenic mice. *Curr. Protoc. Neurosci.* Chapter 5 Molecular Neuroscience, 5.21.1–5.21.13.

Gong, S., Yang, X. W., Li, C., and Heintz, N. (2002). Highly efficient modification of bacterial artificial chromosomes (BACs) using novel shuttle vectors containing the R6Kgamma origin of replication. *Genome Res.* **12,** 1992–1998.

Gray, M., Shirasaki, D. I., Cepeda, C., Andre, V. M., Wilburn, B., Lu, X. H., Tao, J., Yamazaki, I., Li, S. H., Sun, Y. E., Li, X. J., Levine, M. S., *et al.* (2008). Full-length human mutant huntingtin with a stable polyglutamine repeat can elicit progressive and selective neuropathogenesis in BACHD mice. *J. Neurosci.* **28,** 6182–6195.

Graybiel, A. M. (2000). The basal ganglia. *Curr. Biol.* **10,** R509–R511.

Hardy, J., Cai, H., Cookson, M. R., Gwinn-Hardy, K., and Singleton, A. (2006). Genetics of Parkinson's disease and parkinsonism. *Ann. Neurol.* **60,** 389–398.

Hattori, N., Matsumine, H., Asakawa, S., Kitada, T., Yoshino, H., Elibol, B., Brookes, A. J., Yamamura, Y., Kobayashi, T., Wang, M., Yoritaka, A., Minoshima, S., *et al.* (1998). Point mutations (Thr240Arg and Gln311Stop) [correction of Thr240Arg and Ala311Stop] in the Parkin gene. *Biochem. Biophys. Res. Commun.* **249,** 754–758.

Heintz, N. (2001). BAC to the future: The use of bac transgenic mice for neuroscience research. *Nat. Rev. Neurosci.* **2,** 861–870.

Hattori, N., and Mizuno, Y. (2004). Pathogenetic mechanisms of parkin in Parkinson's disease. *Lancet* **364,** 722–724

Heintz, N. (2004). Gene expression nervous system atlas (GENSAT). *Nat. Neurosci.* **7,** 483.

Henn, I. H., Gostner, J. M., Lackner, P., Tatzelt, J., and Winklhofer, K. F. (2005). Pathogenic mutations inactivate parkin by distinct mechanisms. *J. Neurochem.* **92,** 114–122.

Hilker, R., Klein, C., Ghaemi, M., Kis, B., Strotmann, T., Ozelius, L. J., Lenz, O., Vieregge, P., Herholz, K., Heiss, W. D., and Pramstaller, P. P. (2001). Positron emission tomographic analysis of the nigrostriatal dopaminergic system in familial parkinsonism associated with mutations in the parkin gene. *Ann. Neurol.* **49,** 367–376.

Hodgson, J. G., Agopyan, N., Gutekunst, C. A., Leavitt, B. R., LePiane, F., Singaraja, R., Smith, D. J., Bissada, N., McCutcheon, K., Nasir, J., Jamot, L., Li, X. J., *et al.* (1999). A YAC mouse model for Huntington's disease with full-length mutant huntingtin, cytoplasmic toxicity, and selective striatal neurodegeneration. *Neuron* **23,** 181–192.

Hyun, D. H., Lee, M., Hattori, N., Kubo, S., Mizuno, Y., Halliwell, B., and Jenner, P. (2002). Effect of wild-type or mutant Parkin on oxidative damage, nitric oxide, antioxidant defenses, and the proteasome. *J. Biol. Chem.* **277,** 28572–28577.

Hyun, D. H., Lee, M., Halliwell, B., and Jenner, P. (2005). Effect of overexpression of wild-type or mutant parkin on the cellular response induced by toxic insults. *J. Neurosci. Res.* **82,** 232–244.

Imarisio, S., Carmichael, J., Korolchuk, V., Chen, C. W., Saiki, S., Rose, C., Krishna, G., Davies, J. E., Ttofi, E., Underwood, B. R., and Rubinsztein, D. C. (2008). Huntington's disease: From pathology and genetics to potential therapies. *Biochem. J.* **412,** 191–209.

Itier, J. M., Ibanez, P., Mena, M. A., Abbas, N., Cohen-Salmon, C., Bohme, G. A., Laville, M., Pratt, J., Corti, O., Pradier, L., Ret, G., Joubert, C., *et al.* (2003). Parkin gene inactivation alters behaviour and dopamine neurotransmission in the mouse. *Hum. Mol. Genet.* **12,** 2277–2291.

Jaenisch, R. (1977). Germ line integration of moloney leukemia virus: Effect of homozygosity at the m-mulV locus. *Cell* **12,** 691–696.

Jakobovits, A., Moore, A. L., Green, L. L., Vergara, G. J., Maynard-Currie, C. E., Austin, H. A., and Klapholz, S. (1993). Germ-line transmission and expression of a human-derived yeast artificial chromosome. *Nature* **362,** 255–258.

Kahle, P. J., Neumann, M., Ozmen, L., and Haass, C. (2000). Physiology and pathophysiology of alpha-synuclein. Cell culture and transgenic animal models based on a Parkinson's disease-associated protein. *Ann. N Y Acad. Sci.* **920,** 33–41.

Khan, N. L., Scherfler, C., Graham, E., Bhatia, K. P., Quinn, N., Lees, A. J., Brooks, D. J., Wood, N. W., and Piccini, P. (2005). Dopaminergic dysfunction in unrelated, asymptomatic carriers of a single parkin mutation. *Neurology* **64,** 134–136.

Klein, C., Lohmann-Hedrich, K., Rogaeva, E., Schlossmacher, M. G., and Lang, A. E. (2007). Deciphering the role of heterozygous mutations in genes associated with parkinsonism. *Lancet Neurol.* **6,** 652–662.

Kobayashi, T., Matsumine, H., Zhang, J., Imamichi, Y., Mizuno, Y., and Hattori, N. (2003). Pseudo-autosomal dominant inheritance of PARK2: Two families with parkin gene mutations. *J. Neurol. Sci.* **207,** 11–17.

Lee, E. C., Yu, D., Martinez de Velasco, J., Tessarollo, L., Swing, D. A., Court, D. L., Jenkins, N. A., and Copeland, N. G. (2001). A highly efficient *Escherichia coli*-based chromosome engineering system adapted for recombinogenic targeting and subcloning of BAC DNA. *Genomics* **73,** 56–65.

Li, Y., Liu, W., Oo, T. F., Wang, L., Tang, Y., Jackson-Lewis, V., Zhou, C., Geghman, K., Bogdanov, M., Przedborski, S., Beal, M. F., Burke, R. E., *et al.* (2009). Mutant LRRK2

(R1441G) BAC transgenic mice recapitulate cardinal features of Parkinson's disease. *Nat. Neurosci.* **12,** 826–828.

Lu, X. H., Fleming, S. M., Meurers, B., Ackerson, L. C., Mortazavi, F., Lo, V., Hernandez, D., Sulzer, D., Jackson, G. R., Maidment, N. T., Chesselet, M. F., and Yang, X. W. (2009). Bacterial artificial chromosome transgenic mice expressing a truncated mutant parkin exhibit age-dependent hypokinetic motor deficits, dopaminergic neuron degeneration, and accumulation of proteinase K-resistant alpha-synuclein. *J. Neurosci.* **29,** 1962–1976.

Mangiarini, L., Sathasivam, K., Seller, M., Cozens, B., Harper, A., Hetherington, C., Lawton, M., Trottier, Y., Lehrach, H., Davies, S. W., and Bates, G. P. (1996). Exon 1 of the HD gene with an expanded CAG repeat is sufficient to cause a progressive neurological phenotype in transgenic mice. *Cell* **87,** 493–506.

Masliah, E., Rockenstein, E., Veinbergs, I., Mallory, M., Hashimoto, M., Takeda, A., Sagara, Y., Sisk, A., and Mucke, L. (2000). Dopaminergic loss and inclusion body formation in alpha-synuclein mice: Implications for neurodegenerative disorders. *Science* **287,** 1265–1269.

Matsuoka, Y., Vila, M., Lincoln, S., McCormack, A., Picciano, M., LaFrancois, J., Yu, X., Dickson, D., Langston, W. J., McGowan, E., Farrer, M., Hardy, J., *et al.* (2001). Lack of nigral pathology in transgenic mice expressing human alpha-synuclein driven by the tyrosine hydroxylase promoter. *Neurobiol. Dis.* **8,** 535–539.

Menalled, L. B., and Chesselet, M. F. (2002). Mouse models of Huntington's disease. *Trends Pharmacol. Sci.* **23,** 32–39.

Morris, H. R. (2005). Genetics of Parkinson's disease. *Ann. Med.* **37,** 86–96.

Palacino, J. J., Sagi, D., Goldberg, M. S., Krauss, S., Motz, C., Wacker, M., Klose, J., and Shen, J. (2004). Mitochondrial dysfunction and oxidative damage in parkin-deficient mice. *J. Biol. Chem.* **279,** 18614–18622

Pearson, B. E., and Choi, T. K. (1993). Expression of the human beta-amyloid precursor protein gene from a yeast artificial chromosome in transgenic mice. *Proc. Natl. Acad. Sci. USA* **90,** 10578–10582.

Perez, F. A., and Palmiter, R. D. (2005). Parkin-deficient mice are not a robust model of parkinsonism. *Proc. Natl. Acad. Sci. USA.* **102,** 2174–2179.

Peterson, K. R., Clegg, C. H., Li, Q., and Stamatoyannopoulos, G. (1997). Production of transgenic mice with yeast artificial chromosomes. *Trends Genet.* **13,** 61–66.

Polymeropoulos, M. H., Lavedan, C., Leroy, E., Ide, S. E., Dehejia, A., Dutra, A., Pike, B., Root, H., Rubenstein, J., Boyer, R., Stenroos, E. S., Chandrasekharappa, S., *et al.* (1997). Mutation in the alpha-synuclein gene identified in families with Parkinson's disease. *Science* **276,** 2045–2047.

Rao, S. S., Hofmann, L. A., and Shakil, A. (2006). Parkinson's disease: Diagnosis and treatment. *Am. Fam. Physician.* **74,** 2046–2054.

Richfield, E. K., Thiruchelvam, M. J., Cory-Slechta, D. A., Wuertzer, C., Gainetdinov, R. R., Caron, M. G., Di Monte, D. A., and Federoff, H. J. (2002). Behavioral and neurochemical effects of wild-type and mutated human alpha-synuclein in transgenic mice. *Exp. Neurol.* **175,** 35–48.

Ristevski, S. (2005). Making better transgenic models: Conditional, temporal, and spatial approaches. *Mol. Biotechnol.* **29,** 153–163.

Ross, C. A. (2004). Huntington's disease: New paths to pathogenesis. *Cell* **118,** 4–7.

Sang, T. K., Chang, H. Y., Lawless, G. M., Ratnaparkhi, A., Mee, L., Ackerson, L. C., Maidment, N. T., Krantz, D. E., and Jackson, G. R. (2007). A *Drosophila* model of mutant human parkin-induced toxicity demonstrates selective loss of dopaminergic neurons and dependence on cellular dopamine. *J. Neurosci.* **27,** 981–992.

Sharp, A. H., Loev, S. J., Schilling, G., Li, S. H., Li, X. J., Bao, J., Wagster, M. V., Kotzuk, J. A., Steiner, J. P., Lo, A., Hedreen, J., Sisodia, S., *et al.* (1995). Widespread expression of Huntington's disease gene (IT15) protein product. *Neuron* **14,** 1065–1074.

Shen, J. (2004). Protein kinases linked to the pathogenesis of Parkinson's disease. *Neuron* **44,** 575–577.

Singleton, A. B., Farrer, M., Johnson, J., Singleton, A., Hague, S., Kachergus, J., Hulihan, M., Peuralinna, T., Dutra, A., Nussbaum, R., Lincoln, S., Crawley, A., *et al.* (2003). alpha-Synuclein locus triplication causes Parkinson's disease. *Science* **302,** 841.

Slow, E. J., van Raamsdonk, J., Rogers, D., Coleman, S. H., Graham, R. K., Deng, Y., Oh, R., Bissada, N., Hossain, S. M., Yang, Y. Z., Li, X. J., Simpson, E. M., *et al.* (2003). Selective striatal neuronal loss in a YAC128 mouse model of Huntington disease. *Hum. Mol. Genet.* **12,** 1555–1567.

Sopher, B. L., and La Spada, A. R. (2006). Efficient recombination-based methods for bacterial artificial chromosome fusion and mutagenesis. *Gene* **371,** 136–143.

Sopher, B. L., Thomas, P. S. Jr., LaFevre-Bernt, M. A., Holm, I. E., Wilke, S. A., Ware, C. B., Jin, L. W., Libby, R. T., Ellerby, L. M., and La Spada, A. R. (2004). Androgen receptor YAC transgenic mice recapitulate SBMA motor neuronopathy and implicate VEGF164 in the motor neuron degeneration. *Neuron* **41,** 687–699.

Spillantini, M. G., Schmidt, M. L., Lee, V. M., Trojanowski, J. Q., Jakes, R., and Goedert, M. (1997). Alpha-synuclein in Lewy bodies. *Nature* **388,** 839–840.

Springer, W., Hoppe, T., Schmidt, E., and Baumeister, R. (2005). A *Caenorhabditis elegans* Parkin mutant with altered solubility couples alpha-synuclein aggregation to proteotoxic stress. *Hum. Mol. Genet.* **14,** 3407–3423.

Sriram, S. R., Li, X., Ko, H. S., Chung, K. K., Wong, E., Lim, K. L., Dawson, V. L., and Dawson, T. M. (2005). Familial-associated mutations differentially disrupt the solubility, localization, binding and ubiquitination properties of parkin. *Hum. Mol. Genet.* **14,** 2571–2586.

Sun, M., Latourelle, J. C., Wooten, G. F., Lew, M. F., Klein, C., Shill, H. A., Golbe, L. I., Mark, M. H., Racette, B. A., Perlmutter, J. S., Parsian, A., Guttman, M., *et al.* (2006). Influence of heterozygosity for parkin mutation on onset age in familial Parkinson disease: The GenePD study. *Arch. Neurol.* **63,** 826–832.

Thomas, B., and Beal, M. F. (2007). Parkinson's disease. *Hum. Mol. Genet.* **16**(Spec No. 2), R183–R194.

Trojanowski, J. Q., and Lee, V. M. (1998). Aggregation of neurofilament and alpha-synuclein proteins in Lewy bodies: Implications for the pathogenesis of Parkinson disease and Lewy body dementia. *Arch. Neurol.* **55,** 151–152.

van der Putten, H., Wiederhold, K. H., Probst, A., Barbieri, S., Mistl, C., Danner, S., Kauffmann, S., Hofele, K., Spooren, W. P., Ruegg, M. A., Lin, S., Caroni, P., *et al.* (2000). Neuropathology in mice expressing human alpha-synuclein. *J. Neurosci.* **20,** 6021–6029.

Van Keuren, M. L., Gavrilina, G. B., Filipiak, W. E., Zeidler, M. G., and Saunders, T. L. (2009). Generating transgenic mice from bacterial artificial chromosomes: Transgenesis efficiency, integration and expression outcomes. *Transgenic Res.* **18,** 769–785.

Van Raamsdonk, J. M., Murphy, Z., Slow, E. J., Leavitt, B. R., and Hayden, M. R. (2005). Selective degeneration and nuclear localization of mutant huntingtin in the YAC128 mouse model of Huntington disease. *Hum. Mol. Genet.* **14,** 3823–3835.

Varrone, A., Pellecchia, M. T., Amboni, M., Sansone, V., Salvatore, E., Ghezzi, D., Garavaglia, B., Brice, A., Brunetti, A., Bonavita, V., De Michele, G., Salvatore, M., *et al.* (2004). Imaging of dopaminergic dysfunction with [123I]FP-CIT SPECT in early-onset parkin disease. *Neurology* **63,** 2097–2103.

von Coelln, R., Dawson, V. L., and Dawson, T. M. (2004). Parkin-associated Parkinson's disease. *Cell Tissue Res.* **318,** 175–184.

Vonsattel, J. P., and DiFiglia, M. (1998). Huntington disease. *J. Neuropathol. Exp. Neurol.* **57,** 369–384.

Wang, C., Ko, H. S., Thomas, B., Tsang, F., Chew, K. C., Tay, S. P., Ho, M. W., Lim, T. M., Soong, T. W., Pletnikova, O., Troncoso, J., Dawson, V. L., *et al.* (2005a). Stress-induced alterations in parkin solubility promote parkin aggregation and compromise parkin's protective function. *Hum. Mol. Genet.* **14,** 3885–3897.

Wang, C., Tan, J. M., Ho, M. W., Zaiden, N., Wong, S. H., Chew, C. L., Eng, P. W., Lim, T. M., Dawson, T. M., and Lim, K. L. (2005b). Alterations in the solubility and intracellular localization of parkin by several familial Parkinson's disease-linked point mutations. *J. Neurochem.* **93,** 422–431.

Wang, C., Lu, R., Ouyang, X., Ho, M. W., Chia, W., Yu, F., and Lim, K. L. (2007). *Drosophila* overexpressing parkin R275W mutant exhibits dopaminergic neuron degeneration and mitochondrial abnormalities. *J. Neurosci.* **27,** 8563–8570.

Warming, S., Costantino, N., Court, D. L., Jenkins, N. A., and Copeland, N. G. (2005). Simple and highly efficient BAC recombineering using galK selection. *Nucleic. Acids. Res.* **33,** e36

Yang, X. W., and Gong, S. (2005). An overview on the generation of BAC transgenic mice for neuroscience research. *Curr. Protoc. Neurosci.* Chapter 5: Unit 5.20.

Yang, X. L., and Lu, X. H. (2008). The BAC transgenic approach to study Parkinson's disease in mice. *In* "Parkinson's Disease: Pathogenic and Therapeutic Insights from Toxin and Genetic Models", (S. P. Richard Nass, Ed.), Elsevier, San Diego, CA.

Yang, X. W., Model, P., and Heintz, N. (1997). Homologous recombination based modification in *Escherichia coli* and germline transmission in transgenic mice of a bacterial artificial chromosome. *Nat. Biotechnol.* **15,** 859–865.

Zhang, Y., Buchholz, F., Muyrers, J. P., and Stewart, A. F. (1998). A new logic for DNA engineering using recombination in *Escherichia coli. Nat. Genet.* **20,** 123–128.

Zimprich, A., Biskup, S., Leitner, P., Lichtner, P., Farrer, M., Lincoln, S., Kachergus, J., Hulihan, M., Uitti, R. J., Calne, D. B., Stoessl, A. J., Pfeiffer, R. F., *et al.* (2004). Mutations in LRRK2 cause autosomal-dominant parkinsonism with pleomorphic pathology. *Neuron* **44,** 601–607.

Zoghbi, H. Y., and Orr, H. T. (2000). Glutamine repeats and neurodegeneration. *Annu. Rev. Neurosci.* **23,** 217–247.

BEHAVIORAL OUTCOME MEASURES FOR THE ASSESSMENT OF SENSORIMOTOR FUNCTION IN ANIMAL MODELS OF MOVEMENT DISORDERS

Sheila M. Fleming

Departments of Psychology and Neurology, University of Cincinnati, Cincinnati, Ohio 45221, USA

Animal models have and continue to contribute to our understanding of the neurobiology many types of disorders. In movement disorders such as Parkinson's disease (PD), animal models have directly led to various therapeutic treatments such as deep brain stimulation. To facilitate the development of potential therapeutics, sensitive and reliable outcome measures in animal models are necessary to maximize their benefit. In this chapter, behavioral outcome measures, sensitive to varying degrees of sensorimotor dysfunction, are reviewed in rats and mice.

I. Introduction

The relationship between dopamine (DA) and movement has been studied in depth since the discovery of DA in the brain during the late 1950s and its association with the disorder Parkinson's disease (PD) soon thereafter. In PD, DA neurons in the substantia nigra progressively degenerate leading to motor abnormalities including resting tremor, bradykinesia, rigidity, and gait

INTERNATIONAL REVIEW OF NEUROBIOLOGY, VOL. 89
DOI: 10.1016/S0074-7742(09)89003-X

disturbances. A number of neurotoxins have been used to model the loss of nigrostriatal DA neurons seen in PD, the most extensively studied model being the 6-hydroxydopamine (6-OHDA) rat model. More recently, the discovery of genetic forms of PD has led to a new generation of novel genetic mouse models with the most detailed behavioral characterization performed in mice that overexpress the presynaptic protein alpha synuclein. In both models sensorimotor tests that are sensitive to dysfunction and loss of nigrostriatal DA neurons have been developed and are important tools in providing endpoint measures for preclinical testing of potential therapeutic treatments for PD. Although the tests described in this chapter have been used in models of PD, they are also applicable to other types of movement disorders such as stroke and spinal cord injury (Schallert *et al.*, 2000).

II. Sensorimotor Tests for Unilateral 6-OHDA Rat

The toxin 6-OHDA is widely used to create animal models of PD (Kostrzewa and Jacobwitz, 1974; Schallert and Wilcox, 1985; Ungerstedt, 1968, 1971). 6-OHDA does not cross the blood–brain barrier and therefore is injected directly into the substantia nigra. Once in the cell, 6-OHDA forms cytotoxic products like hydrogen peroxide (Heikkila and Cohen, 1971), superoxide, and hydroxy radicals (Cohen and Heikkila, 1974; Heikkila and Cohen, 1973), leading to cell death.

Rats with unilateral DA depletions display parkinson-like symptoms on only one side of the body, they show asymmetrical sensorimotor impairments of the limbs that are contralateral to the side of the lesion such as akinesia, somatosensory neglect, and postural abnormalities (Schallert and Tillerson, 2000; Schallert *et al.*, 1982, 1992). Here, four commonly used tests that are sensitive to asymmetrical impairments are briefly described.

A. Limb-Use Asymmetry

Forelimb use during exploratory activity is an elegant and simple test to measure limb use asymmetry. The rat is placed in a clear cylinder (20 cm diameter and 30 cm height) for 3–10 min with a mirror positioned at an angle to enable a complete view of the rat at any place in the cylinder. The cylindrical shape encourages vertical exploration of the walls with the forelimbs, as well as landing activity. Simply, individual and combined limb use is measured. Rats with a unilateral 6-OHDA lesion prefer to use the limb ipsilateral to the lesion. This test is sensitive to varying degrees of nigrostriatal DA neuron loss and has been used extensively to evaluate the efficacy of various types of transplants and viral vectors (Connor *et al.*, 1999; Kozlowski *et al.*, 2000; Luo *et al.*, 2002; Yang *et al.*, 2002).

B. Movement Initiation

To test forelimb movement initiation, stepping movements made with the ipsilateral and contralateral forelimbs are assessed using an isolated forelimb akinesia test (Lindner *et al.*, 1996; Olsson *et al.*, 1995; Schallert *et al.*, 1992, 2000; Tillerson *et al.*, 2001, 2002a). Here, the rat is held by its torso with its hindlimbs and one forelimb lifted above the surface of a table so that the weight of its body is supported by one forelimb alone. The number of self-initiated steps made in a 10-s trial can be recorded for each forelimb for multiple trials and then averaged. To account for individual differences in the number of steps made between animals, ipsilateral minus contralateral scores can be calculated.

C. Adjusting Steps

The step test is used to measure postural stability (Chang *et al.*, 1999; Fleming *et al.*, 2004b; Schallert and Tillerson, 2000). In this test, animals are held in the same manner as in the movement initiation test where one forelimb bears the weight of the animal. The animal is then moved laterally across a distance of 80 cm on a tabletop. The number of adjusting steps made as the animal is moved across a table is recorded for each forelimb. The average number of steps in three trials for each forelimb can be used for analysis. Animals with nigrostriatal damage typically drag their forelimb across the tabletop instead of making adjusting steps (Schallert *et al.*, 1978).

D. Somatosensory Neglect

Somatosensory asymmetry is assessed using a bilateral tactile stimulation test (Fleming *et al.*, 2005; Schallert *et al.*, 1982, 1983). Animals are first tested to indicate the presence of a somatosensory asymmetry. This is done by removing the animal from the home cage and attaching adhesive stimuli (Avery adhesive-backed labels, 113 mm^2) to the distal–radial aspect of each forelimb in random order. After being returned to the home cage, rats contact and remove the stimuli one at a time. The order and latency of stimulus contact and removal is recorded for each of five trials. The order of contact is used to determine whether animals show a bias for the stimulus on the forelimb unaffected by the injury.

III. Sensorimotor Tests for Genetic Mouse Models

Within the past decade several genetic mutations causing rare familial cases of PD have been identified. Mutations in the presynaptic protein alpha synuclein were some of the first mutations to be described (Chartier-Harlin *et al.*, 2004; Kruger *et al.*, 1998; Polymeropoulos *et al.*, 1997; Singleton *et al.*, 2003). Soon after this, genetic mice carrying similar mutations were generated. In addition to mice with mutations associated with familial forms of PD, there have also been mice generated that have mutations that interfere with the development of nigrostriatal DA neurons (Hwang *et al.*, 2003). The nigrostriatal system is altered to different degrees in these mice therefore sensitive and reliable sensorimotor tests like those described for the unilateral 6-OHDA rat have been developed to detect bilateral deficits in mice.

A. Challenging Beam Traversal

Following injury or disease, both humans and animals use compensatory strategies to perform tasks accurately, making it difficult to detect impairments, especially in the early stages of the disease (LeVere, 1988; Schallert and Hall, 1988; Whishaw, 2000). Therefore, it is important to challenge the animals to the limit of their abilities to uncover early effects of the mutations. Motor performance and coordination can be measured with the challenging beam traversal test (Fleming *et al.*, 2004a, 2006; Hwang *et al.*, 2005; Lu *et al.*, 2009). Briefly, the beam consists of four sections (25 cm each, 1 m total length), each section having a different width. The beam starts at a width of 3.5 cm and gradually narrows to 0.5 cm × 1 cm increments. Animals are trained to traverse the length of the beam starting at the widest section and ending at the narrowest, most difficult, section. On the day of the test, a mesh grid (1 cm^2) of corresponding width is placed over the beam surface leaving approximately a one cm space between the grid and the beam surface, which serves as a crutch to prevent compensatory motor learning that can otherwise mask extant deficits (Schallert *et al.*, 2002). Animals are then videotaped while traversing the grid-surfaced beam. Videotapes are rated for errors, number of steps made by each animal, and time to traverse. This test has been shown to be highly sensitive in mice with mutations associated with familial PD (Fleming *et al.*, 2004a; Goldberg *et al.*, 2003; Lu *et al.*, 2009) and in mice with a developmental loss of nigrostriatal DA neurons (Hwang *et al.*, 2005). In addition impairments in mice with DA cell loss can be reversed with L-DOPA (Hwang *et al.*, 2005).

B. Pole Test and Inverted Grid

The pole and inverted grid tests are also good measures of motor performance and coordination. For the inverted grid test, animals are placed upside down on a grid above the ground. For the pole test, animals are placed head up on the top of a pole and time to orient the body downward as well as time to descend are measured. Both tests are sensitive measures in alpha synuclein overexpressing mice (Fleming *et al.*, 2004a) and mice with a loss of DA neurons (Hwang *et al.*, 2005; Tillerson *et al.*, 2002b).

C. Response to Sensory Stimuli

Similar to the test of somatosensory asymmetry in rats, response to sensory stimuli can be measured in mice. In this test, small adhesive stimuli (Avery adhesive-backed labels, 1/4″ round) are placed on the snout of the mouse and the time to make contact and remove the stimulus are recorded. If the animal does not remove the stimulus within 60 s then the experimenter removes it, and the trial for the next mouse is initiated. Stimulus contact and removal times are calculated for each animal. This test has been shown to be sensitive in many genetic mouse models of PD including mice that overexpress alpha synuclein (Fleming *et al.*, 2004a), DJ-1 knockout mice (Chen *et al.*, 2005), parkin knockout mice (Goldberg *et al.*, 2003), and most recently mice with a parkin Q113X mutation (Lu *et al.*, 2009).

D. Spontaneous Activity

Adapted from the test of limb use asymmetry, spontaneous movement was measured by placing animals in a small transparent cylinder (height, 15.5 cm; diameter, 12.7 cm; Fleming *et al.*, 2004a, 2006; Hwang *et al.*, 2005; Lu *et al.*, 2009). The cylinder was placed on a piece of glass with a mirror was positioned at an angle beneath the cylinder to allow a clear view of motor movements along the ground and walls of the cylinder. The number of rears, forelimb and hindlimb steps, and time spent grooming were measured. Videotapes were viewed and rated in slow motion by an experimenter blind to the mouse genotype. Number of steps, rears and time spent grooming are measured.

In this test, alpha synuclein overexpressing mice (Rockenstein *et al.*, 2002) and parkin Q311X mice (Lu *et al.*, 2009) showed significant reductions in spontaneous activity that persist over time (Fleming *et al.*, 2004a). In addition, mice with a developmental loss of DA neurons also show a reduction in activity that is reversed by L-DOPA (Hwang *et al.*, 2005).

E. Nest Building

Nest building is a natural mouse behavior related to thermoregulation and pup survival (Broida and Svare, 1982; Crawley, 2000; Lynch, 1980). Analysis of nest building behavior has been used to assess nigrostriatal sensorimotor function in rodents (Hofele *et al.*, 2001; Sedelis *et al.*, 2001; Szczypka *et al.*, 2001; Upchurch and Schallert, 1983). In this test, cotton material for nest building is weighed and then placed in the feeder bin of the animal's home cage. By placing the nesting material in the feeder bin of the cage, animals must rear up and pull the nesting material from the feeder making it a more challenging test than if the nesting material was just placed on the floor of the cage. The amount of cotton used is measured after a 24-h period. A control test with nesting material in the cage is conducted to rule out a decrease in nest building motivation.

All of the tests described in this chapter reflect measures that are sensitive to varying levels of dysfunction within sensorimotor systems. In addition to models of Parkinsonism tests such as limb use asymmetry and somatosensory asymmetry are often used in cortical lesion models (Schallert *et al.*, 2000). In mice, the challenging beam has also been used in a mouse model of ataxia telangiectasia and a model of familial amyloidosis (Page *et al.*, 2009; Reliene *et al.*, 2009) highlighting the broad use of these outcome measures.

Acknowledgements

This work was funded by American Parkinson Disease Association and the Chen Family.

References

Broida, J., and Svare, B. (1982). Strain-typical patterns of pregnancy-induced nestbuilding in mice: Maternal and experiential influences. *Physiol. Behav.* **29**(1), 53–57.

Chang, J. W., Wachtel, S. R., Young, D., and Kang, U. J. (1999). Biochemical and anatomical characterization of forepaw adjusting steps in rat models of Parkinson's disease: Studies on medial forebrain bundle and striatal lesions. *Neuroscience* **88**(2), 617–628.

Chartier-Harlin, M. C., Kachergus, J., Roumier, C., Mouroux, V., Douay, X., Lincoln, S., Levecque, C., Larvor, L., Andrieux, J., Hulihan, M., Waucquier, N., Defebvre, L., *et al.* (2004). Alpha-synuclein locus duplication as a cause of familial Parkinson's disease. *Lancet* **364,** 1167–1169.

Chen, L., Cagniard, B., Matthews, T., Jones, S., Koh, H. C., Ding, Y., Carvey, P. M., Ling, Z., Kang, U. J., and Zhuang, X. (2005). Age-dependent motor deficits and dopaminergic dysfunction in DJ-1 null mice. *J. Biol. Chem.* **280**(22), 21418–21426.

Cohen, G., and Heikkila, R. E. (1974). The generation of hydrogen peroxide, superoxide radical, and hydroxyl radical by 6-hydroxydopamine, dialuric acid, and related cytotoxic agents. *J. Biol. Chem.* **249**(8), 2447–2452.

Connor, B., Kozlowski, D. A., Schallert, T., Tillerson, J. L., Davidson, B. L., and Bohn, M. C. (1999). Differential effects of glial cell line-derived neurotrophic factor (GDNF) in the striatum and substantia nigra of the aged Parkinsonian rat. *Gene Ther.* **6**(12), 1936–1951.

Crawley, J. N. (2000). What's Wrong With My Mouse? Behavioral Phenotyping of Transgenic and Knockout Mice Wiley-Liss, New York.

Fleming, S. M., Salcedo, J., Fernagut, P. O., Rockenstein, E., Masliah, E., Levine, M. S., and Chesselet, M.-F. (2004a). Early and progressive sensorimotor anomalies in mice overexpressing wild-type human alpha-synuclein. *J. Neurosci.* **24,** 9434–9440.

Fleming, S. M., Zhu, C., Fernagut, P.-O., Mehta, A., DiCarlo, C. D., Seaman, R. L., and Chesselet, M.-F. (2004b). Behavioral and immunohistochemical effects of chronic intravenous and subcutaneous infusions of varying doses of rotenone. *Exp. Neurol.* **187,** 418–429.

Fleming, S. M., Delville, Y., and Schallert, T. (2005). An intermittent, controlled-rate, slow progressive degeneration model of Parkinson's disease: Antiparkinson effects of sinemet and protective effects of methylphenidate. *Behav. Brain Res.* **156**(2), 201–213.

Fleming, S. M., Salcedo, J., Hutson, C. B., Rockenstein, E., Masliah, E., Levine, M. S., and Chesselet, M.-F. (2006). Behavioral effects of dopaminergic agonists in transgenic mice overexpressing human wildtype alpha-synuclein. *Neuroscience* **142**(4), 1245–1253.

Goldberg, M. S., Fleming, S. M., Palacino, J. J., Cepeda, C., Lam, H. A., Bhatnagar, A., Meloni, E. G., Wu, N., Ackerson, L. C., Klapstein, G. J., Gajendiran, M., Roth, B. L., *et al.* (2003). Parkin-deficient mice exhibit nigrostriatal deficits but not loss of dopaminergic neurons. *J. Biol. Chem.* **278,** 43628–43635.

Heikkila, R., and Cohen, G. (1971). Inhibition of biogenic amine uptake by hydrogen peroxide: A mechanism for toxic effects of 6-hydroxydopamine. *Science* **172**(989), 1257–1258.

Heikkila, R., and Cohen, G. (1973). 6-Hydroxydopamine: Evidence for superoxide radical as an oxidative intermediate. *Science* **181**(98), 456–457.

Hofele, K., Sedelis, M., Auburger, G. W., Morgan, S., Huston, J. P., and Schwarting, R. K. (2001). Evidence for a dissociation between MPTP toxicity and tyrosinase activity based on congenic mouse strain susceptibility. *Exp. Neurol.* **168**(1), 116–122.

Hwang, D. Y., Ardayfio, P., Kang, U. J., Semina, E. V., and Kim, K. S. (2003). Selective loss of dopaminergic neurons in the substantia nigra of Pitx3-deficient aphakia mice. *Brain Res. Mol. Brain Res.* **114**(2), 123–131.

Hwang, D. Y., Fleming, S. M., Ardayfio, P., Moran-Gates, T., Kim, H., Tarazi, F. I., Chesselet, M- F., and Kim, K. S. (2005). 3,4-Dihydroxyphenylalanine reverses the motor deficits in Pitx3-deficient aphakia mice: Behavioral characterization of a novel genetic model of Parkinson's disease. *J. Neurosci.* **25**(8), 2132–2137.

Kostrzewa, R. M., and Jacobwitz, D. M. (1974). Pharmacological actions of 6-hydroxydopamine. *Pharmacol. Rev.* **26**(3), 199–288.

Kozlowski, D. A., Connor, B., Tillerson, J. L., Schallert, T., and Bohn, M. C. (2000). Delivery of a GDNF gene into the substantia nigra after a progressive 6-OHDA lesion maintains functional nigrostriatal connections. *Exp. Neurol.* **166**(1), 1–15.

Kruger, R., Kuhn, W., Muller, T., Woitalla, D., Graeber, M., Kosel, S., Przuntek, H., Epplen, J. T., Schöls, L., and Riess, O. (1998). Ala30Pro mutation in the gene encoding alpha-synuclein in Parkinson's disease. *Nat. Genet.* **18,** 106–108.

LeVere, T. E. (1988). Neural system imbalances and the consequences of large brain injuries. *In* "Brain Injury and Recovery, Theoretical and Controversial Issues," (S. Finger, T. E. LeVere, C. R. Almi, and D. G. Stein, Eds.), pp. 1–362. Plenum Press, New York.

Lindner, M. D., Winn, S. R., Baetge, E. E., Hammang, J. P., Gentile, F. T., Doherty, E., McDermott, P. E., Frydel, B., Ullman, M. D., Schallert, T., and Emerich, D. F. (1996). Implantation of encapsulated catecholamine and GDNF-producing cells in rats with unilateral dopamine depletions and parkinsonian symptoms. *Exp. Neurol.* **132**(1), 62–76.

Lu, X. H., Fleming, S. M., Meurers, B., Mortazavi, F., Lo, V., Hernandez, D., Sulzer, D., Jackson, G. R., Maidment, N. T., Chesselet, M.-F., and Yang, X. W. (2009). BAC mice expressing a truncated mutant parkin exhibit progressive motor deficits and late-onset dopaminergic neuron degeneration. *J. Neurosci.* **29**(7), 1962–1976.

Luo, J., Kaplitt, M. G., Fitzsimons, H. L., Zuzga, D. S., Liu, Y., Oshinsky, M. L., and During, M. J. (2002). Subthalamic GAD gene therapy in a Parkinson's disease rat model. *Science* **298**(5592), 425–429.

Lynch, C. B. (1980). Response to divergent selection for nesting behavior in *Mus musculus*. *Genetics* **96**(3), 757–765.

Olsson, M., Nikkhah, G., Bentlage, C., and Bjorklund, A. (1995). Forelimb akinesia in the rat Parkinson model: Differential effects of dopamine agonists and nigral transplants as assessed by a new stepping test. *J. Neurosci.* **15**(5 Pt 2), 3863–3875.

Page, L. J., Suk, J- Y., Bazhenova, L., Fleming, S. M., Wood, M., Jiang, Y., Guo, L. T., Misizin, A. P., Kisilevsky, R., Shelton, G. D., Kelly, J. W., and Balch, W. E. (2009). A gelsolin amyloidosis mouse exhibiting an age-associated decline in proteostasis triggers sporadic inclusion body myositis (sIBM). *PNAS.* **106**(27), 11125–11130.

Polymeropoulos, M. H., Lavedan, C., Leroy, E., Ide, S. E., Dehejia, A., Dutra, A., Pike, B., Root, H., Rubenstein, J., Boyer, R., Stenroos, E. S., Chandrasekharappa, S., *et al.* (1997). Mutation in the alpha-synuclein gene identified in families with Parkinson's disease. *Science* **276,** 2045–2047.

Reliene, R., Fleming, S. M., Chesselet, M- F., and Schiestl, R. H. (2008). Effects of antioxidants on cancer prevention and neuromotor deficits in *ATM* deficient mice. *Food Chem. Toxicol.* **46,** 1371–1377.

Rockenstein, E., Mallory, M., Hashimoto, M., Song, D., Shults, C. W., Lang, I., and Masliah, E. (2002). Differential neuropathological alterations in transgenic mice expressing alpha-synuclein from the platelet-derived growth factor and Thy-1 promoters. *J. Neurosci. Res.* **68,** 568–578.

Schallert, T., and Hall, S. (1988). 'Disengage' sensorimotor deficit following apparent recovery from unilateral dopamine depletion. *Behav. Brain Res.* **30,** 15–24.

Schallert, T., and Tillerson, J. L. (2000). Intervention strategies for degeneration of dopamine neurons in parkinsonism: Optimizing behavioral assessment of outcome. *In* "Central Nervous System Diseases," (D. F. Emerich, R. L. Dean III, and P. R. Sandberg, Eds.), Humana Press, Totowa, NJ.

Schallert, T., and Wilcox, R. E. (1985). Neurotransmitter-selective brain lesions. *In* "Neuromethods (Series 1: Neurochemistry), General Neurochemical Techniques," (A. A. Boulton and G. B. Baker, Eds.), pp. 343–387. Humana Press, Totowa, NJ.

Schallert, T., Whishaw, I. Q., Ramirez, V. D., and Teitelbaum, P. (1978). Compulsive, abnormal walking caused by anticholinergics in akinetic, 6-hydroxydopamine-treated rats. *Science* **199,** 1461–1463.

Schallert, T., Upchurch, M., Lobaugh, N., Farrar, S. B., Spirduso, W. W., Gilliam, P., Vaughn, D., and Wilcox, R. E. (1982). Tactile extinction: Distinguishing between sensorimotor and motor asymmetries in rats with unilateral nigrostriatal damage. *Pharmacol. Biochem. Behav.* **16**(3), 455–462.

Schallert, T., Upchurch, M., Wilcox, R. E., and Vaughn, D. M. (1983). Posture-independent sensorimotor analysis of inter-hemispheric receptor asymmetries in neostriatum. *Pharmacol. Biochem. Behav.* **18**(5), 753–759.

Schallert, T., Norton, D., and Jones, T. A. (1992). A clinically relevant unilateral rat model of parkinsonian akinesia. *J. Neural. Trans. Plasticity* **3,** 332–333.

Schallert, T., Fleming, S. M., Leasure, J. L., Tillerson, J. L., and Bland, S. T. (2000). CNS plasticity and assessment of forelimb sensorimotor outcome in unilateral rat models of stroke, cortical ablation, parkinsonism and spinal cord injury. *Neuropharmacology* **39**(5), 777–787.

Schallert, T., Woodlee, M. T., and Fleming, S. M. (2002). Disentangling multiple types of recovery from brain injury. *In* "Pharmacology of Cerebral Ischemia," (J. Krieglstein and S. Klumpp, Eds.), pp. 201–216. Medpharm Scientific Publishers, Stuttgart, Germany.

Sedelis, M., Schwarting, R. K., and Huston, J. P. (2001). Behavioral phenotyping of the MPTP mouse model of Parkinson's disease. *Behav. Brain Res.* **125,** 109–125.

Singleton, A. B., Farrer, M., Johnson, J., Singleton, A., Hague, S., Kachergus, J., Hulihan, M., Peuralinna, T., Dutra, A., Nussbaum, R., Lincoln, S., Crawley, A., *et al.* (2003). alpha-Synuclein locus triplication causes Parkinson's disease. *Science* **302,** 841.

Szczypka, M. S., Kwok, K., Brot, M. D., Marck, B. T., Matsumoto, A. M., Donahue, B. A., and Palmiter, R. D. (2001). Dopamine production in the caudate putamen restores feeding in dopamine-deficient mice. *Neuron* **30**(3), 819–828.

Tillerson, J. L., Cohen, A. D., Philhower, J., Miller, G. W., Zigmond, M. J., and Schallert, T. (2001). Forced limb-use effects on the behavioral and neurochemical effects of 6-hydroxydopamine. *J. Neurosci.* **21**(12), 4427–4435.

Tillerson, J. L., Cohen, A. D., Caudle, W. M., Zigmond, M. J., Schallert, T., and Miller, G. W. (2002a). Forced nonuse in unilateral parkinsonian rats exacerbates injury. *J. Neurosci.* **22**(15), 6790–6799.

Tillerson, J. L., Caudle, W. M., Reveron, M. E., and Miller, G. W. (2002b). Detection of behavioral impairments correlated to neurochemical deficits in mice treated with moderate doses of 1 methyl-4-phenyl 1,2,3,6-tetrahydropyridine. *Exp. Neurol.* **178,** 80–90.

Ungerstedt, U. (1968). 6-Hydroxy-dopamine induced degeneration of central monoamine neurons. *Eur. J. Pharmacol.* **5**(1), 107–110.

Ungerstedt, U. (1971). Adipsia and aphagia after 6-hydroxydopamine induced degeneration of the nigro-striatal dopamine system. *Act. Physiol. Scand. Suppl.* **367,** 95–122.

Upchurch, M., and Schallert, T. (1983). A behavior analysis of the offspring of "haloperidol-sensitive" and "haloperidol-resistant" gerbils. *Behav. Neural. Biol.* **39**(2), 221–228.

Whishaw, I. Q. (2000). Loss of the innate cortical engram for action patterns used in skilled reaching and the development of behavioral compensation following motor cortex lesions in the rat. *Neuropharmacology* **39**(5), 788–805.

Yang, M., Stull, N. D., Berk, M. A., Snyder, E. Y., and Iacovitti, L. (2002). Neural stem cells spontaneously express dopaminergic traits after transplantation into the intact or 6-hydroxydopamine-lesioned rat. *Exp. Neurol.* **177**(1), 50–60.

THE ROLE OF DNA METHYLATION IN THE CENTRAL NERVOUS SYSTEM AND NEUROPSYCHIATRIC DISORDERS

Jian Feng and Guoping Fan

Department of Human Genetics, David Geffen School of Medicine, University of California, Los Angeles, California 90095, USA

DNA methylation is an epigenetic mechanism in which the methyl group is covalently coupled to the C5 position of the cytosine residue of CpG dinucleotides. DNA methylation generally leads to gene silencing and is catalyzed by a group of enzymes known as DNA methyltransferases (Dnmt). During development, the epigenome undergoes waves of demethylation and methylation changes. As a result, there are cell type/tissue-specific DNA methylation patterns. Since DNA methylation changes only happen during DNA replication to maintain methylation patterns on hemimethylated DNA or establish new methylation, Dnmt expression generally decreases greatly after cell division. However, significant levels of Dnmts were noticed specifically in postmitotic neurons, suggesting a functional importance of Dnmt in the nervous system. Accumulating evidence showed that DNA methylation correlates with certain neuropsychiatric disorders such as schizophrenia, Rett syndrome, and ICF syndrome. Studies of methyl-CpG-binding proteins, Dnmt inhibitors, and Dnmt knockout mice also explored the key role of DNA methylation in neural development, plasticity, learning, and memory. Though an enzyme exhibiting DNA demethylation capability in vertebrates still remains to be identified, DNA methylation status in the CNS appeared to be reversible at certain genomic loci. This supports a maintenance role of Dnmt to prevent active demethylation in postmitotic neurons. Taken together, DNA methylation provides an epigenetic mechanism of gene regulation in neural development, function, and disorders.

INTERNATIONAL REVIEW OF NEUROBIOLOGY, VOL. 89
DOI: 10.1016/S0074-7742(09)89004-1

I. Introduction

The word "epigenetics" is normally defined as "the study of mitotically and/or meiotically heritable changes in gene function that cannot be explained by changes in DNA sequence" (Bird, 2007; Jaenisch and Bird, 2003). However, given the fact that many chromatin marks are short-lived and not transmissible between generations and that DNA methylation pattern can also be rapidly removed during development, arguing for the emphasis on heritability for epigenetics may not be necessary (Bird, 2007).

DNA methylation and histone modification are the two major epigenetic mechanisms. As the fundamental unit of chromatin, the nucleosome consists of DNA wrapping around an octamer histone core. This enables DNA to be tightly packaged into the nucleus. The epigenetic mechanisms adjust gene activity by altering accessibility of DNA to the transcription machinery without changing the genetic code. While modification of DNA by methylation generally leads to gene silencing, posttranslational modifications of histone proteins including acetylation, methylation, phosphorylation, ubiquitination, or sumoylation can lead to both gene activation and repression (Jenuwein and Allis, 2001). One of the best-studied histone modifications is the acetylation status of lysine residues, a reversible process that is catalyzed by either histone acetylase (HAT) or histone deacetylase (HDAC). The addition of an acetyl group decreases the interaction between the negatively charged DNA backbone and the positively charged histone tail. This interaction can lead to a less compact nucleosome, and open access to transcription factor complexes. Conversely, HDAC removes the acetyl group, potentially leading to gene transcription repression. Methylation of histones is more complex since each distinct (mono-, di-, or tri-) methylation of different lysine residue can have opposite effects on transcription. For instance, H3 at lysine 4 (K4) is associated with transcriptional activation whereas methylations on histone H3 at K9 or K27 are usually indicative of transcriptional inhibition (Kouzarides, 2007). Moreover, methylation of histones can be reversed as well (Klose *et al.*, 2006; Shi *et al.*, 2004). This newly discovered mechanism for histone demethylation adds another wrinkle into the understanding of how histones regulate gene expression. It has been demonstrated that epigenetic mechanisms play pivotal roles in translating environmental stimuli into long-lasting gene expression changes in the nervous system which is required under both physiological (such as learning and memory) (Feng *et al.*, 2007) and pathological conditions (such as psychiatric disorders) (Tsankova *et al.*, 2007). However, most of these studies focused on histone modifications. For example, in Aplysia, histone acetylation plays a major role in excitatory transmitter activated gene expression which is needed for long-term synaptic plasticity (Guan *et al.*, 2002). Also, the mutation of a histone acetyltransferase gene is believed to be the cause of a mental retardation disease Rubinstein–Taybi syndrome (RTS) (Petrij *et al.*, 1995). In a histone acetyltransferase mutant RTS mouse model, either suppression of transgene expression or HDAC inhibitor administration could rescue the impairment on long-term memory stabilization (Korzus *et al.*, 2004). More strikingly, in a mouse model of neurodegeneration, increased histone acetylation by inhibitors of HDAC could induce recovery of learning and memory (Fischer *et al.*, 2007; Guan *et al.*, 2009). It is believed that different histone modifications, in combination or alone, define a specific epigenetic mark (histone code) that will lead to different gene expression scenarios. For example, acute and chronic cocaine addiction induced expression of different genes which are associated with different epigenetic regulatory mechanisms (Colvis *et al.*, 2005). The epigenetic mechanism's effect on higher neural functioning appears to be universal since some other epigenetic factors' modulating roles were also found. For instance, KAP1, a vertebrate-specific epigenetic repressor, was found to control gene expression in the hippocampus and modulate the behavioral response to stress (Jakobsson *et al.*, 2008). A further understanding of epigenetic mechanisms in neural functioning may advance our approaches to neuropsychiatric disorder therapies.

Not as dynamic as histone tail modification, covalent 5-cytosine methylation is deemed to be a more static mark. DNA methylation pattern maintenance or establishment only happens during DNA replication (Cameron *et al.*, 1999; Holliday, 1999). DNA methylation normally will not change in nondividing cells which makes DNA methylation a less prominent candidate for dynamic gene expression regulation within postmitotic neurons. However, recent findings support important functional roles of DNA methylation in the nervous system. Since the function of histone modification in the nervous system has been well reviewed (Colvis *et al.*, 2005; Tsankova *et al.*, 2007), we will focus on the role of DNA methylation in this chapter.

II. DNA Methylation and DNA Methyltransferase

Methylation of DNA at the fifth carbon of the cytosine ring exists in all vertebrates, fungi, flowering plants, and some invertebrate insects as well as certain bacterial species. The biological functions of cytosine methylation are quite different in prokaryotes and eukaryotes. In prokaryotes, DNA methylation occurs at adenine as well as cytosine bases, playing a central role in the host restriction system. In eukaryotes, DNA methylation is restricted to cytosine bases and is coupled with a repressed chromatin state and inhibition of gene expression. In mammals, 60–90% of CpG sites are methylated, and the majority of the remaining unmethylated residues exist in CpG islands within gene promoters (Bird, 2002; Jaenisch and Bird, 2003).

DNA methylation is catalyzed by a group of DNA methyltransferases (Dnmts). The eukaryotic Dnmt family has four members: Dnmt1, Dnmt3a, Dnmt3b, and Dnmt3l.

The first murine Dnmt was cloned in 1988 and was later named Dnmt1 (Bestor *et al.*, 1988). It is the most abundant Dnmt in mammalian cells. The enzyme is approximately 1620 amino acid long. The N-terminal 1100 amino acids constitute the regulatory domain and the C-terminal residues constitute the catalytic domain. Dnmt1 predominantly methylates hemimethylated CpG dinucleotides. In a kinetic analysis with unmethylated and hemimethylated 36- and 75-mer oligonucleotides, Dnmt1 showed a 7- to 21-fold preference for hemimethylated DNA (Pradhan *et al.*, 1999). During cell replication, the newly synthesized DNA, which is hemimethylated, must be methylated precisely to maintain the established DNA methylation pattern. Dnmt1 is located at the replication fork and methylates newly biosynthesized DNA strands directly (Hermann *et al.*, 2004). For this reason, Dnmt1 is coined as a maintenance Dnmt.

De novo Dnmts are another group of Dnmts, which can effectively methylate cytosine residues in unmethylated DNA. This event happens particularly during germ cell development and embryogenesis. Dnmt3a and Dnmt3b comprise the two major *de novo* Dnmts. Dnmt3a and Dnmt3b have a high degree of primary structural homology, but they are encoded by different genes mapped to different chromosomes. The Dnmt3 enzymes share similar architectural features to Dnmt1 with a large N-terminal regulatory region connected to a C-terminal catalytic domain (Pradhan and Esteve, 2003). But when functionally compared to Dnmt1, Dnmt3a exhibits a lower level of methyltransferase activity, suggesting that it may require small molecules or protein cofactors for optimal activity (Hsieh, 1999). Overexpression experiments showed that both Dnmt3a and Dnmt3b can methylate a stable episome *in vitro*, but Dnmt3b could not lead to the same degree of *de novo* methylation as Dnmt3a. Dnmt3a and Dnmt3b have differential expression patterns. Dnmt3a is more ubiquitous whereas Dnmt3b is usually expressed at low levels except in the testis, thyroid, and bone marrow (Xie *et al.*, 1999). Although the activities of Dnmt3a and Dnmt3b are reduced after differentiation, Dnmt3b level is greatly increased in tumor cell lines. This evidence suggests that Dnmt3a and Dnmt3b have different functional roles aside from redundancy in *de novo* methylation. Indeed, Dnmt3b is specialized for CpG methylation within repetitive sequences at pericentric satellite regions of chromosomes. Mutation of the Dnmt3b gene was found to be associated with immunodeficiency, centromere instability, and facial anomalies (ICF) syndrome, which is characterized by unmethylated DNA within pericentric regions of chromosomes (Hansen *et al.*, 1999; Xu *et al.*, 1999).

Dnmt3l (Dnmt3 *like*) is the third member of the Dnmt3 family and lacks independent methyltransferases activity. However, Dnmt3l may cooperate with other *de novo* Dnmts. For instance, Dnmt3l binds to the C-terminal of Dnmt3a and

Dnmt3b, increasing the activity of these enzymes up to threefolds (Suetake *et al.*, 2004). In contrast, Dnmt3l has never been found to affect the activity of Dnmt1. Dnmt3l might be important for genomic imprinting since targeted disruption of Dnmt3l resulted in biallelic expression of genes normally imprinted and expressed from one paternal origin (Bourc'his *et al.*, 2001).

Based on sequence homology to Dnmts, a molecule named Dnmt2 was cloned (Goll and Bestor, 2005). It is the most conserved and most widely distributed Dnmt. However, targeted deletion of Dnmt2 in ES cells caused no detectable effect on global DNA methylation, implying that Dnmt2 is not essential for DNA methylation (Okano *et al.*, 1998).

III. DNA Methylation Mediates Gene Silencing

During development, the epigenome undergoes waves of demethylation and methylation change. As a result, various cell type/tissue-specific DNA methylation patterns occur at specific times. Global *de novo* methylation happens specifically during germ cell development and early embryogenesis. How Dnmt enzymes are recruited to the DNA to pursue DNA methylation modification is still unclear. Recent studies suggest there are at least three possible means. First, Dnmt3 enzymes themselves may target DNA via specific domains. For instance, the PWWP domain in the N-terminal half of Dnmt is highly conserved and is important for protein–protein interaction. Mutation of the PWWP domain in *de novo* Dnmts completely abolished its chromatin targeting capacity and led to reduced DNA methylation (Ge *et al.*, 2004). Second, Dnmt may also interact with site-specific transcriptional repressors to target DNA methylation. This idea comes from research on DNA methylation of tumor suppressor genes, which frequently causes transcriptional silencing during cancer. It has been shown that a leukemia-promoting PML-RAR fusion protein induces gene hypermethylation and silencing by recruiting Dnmts to target promoters (Di Croce *et al.*, 2002). Other research also uncovered the mechanistic link between Dnmt and transcriptional repression (Brenner *et al.*, 2005; Robertson *et al.*, 2000). All these findings suggest that the coupling of Dnmt with transcription factors may be a general mechanism for the generation of specific DNA methylation patterns. The third novel mechanism is that DNA methylation might be targeted by transcriptional gene silencing pathways in response to RNA interference (RNAi) signals. In plants, RNAi-mediated transcriptional silencing always results in *de novo* methylation of the gene promoter. Even though the detailed molecular mechanisms have yet to be disclosed, studies of mammalian cell cultures have also suggested a similar mechanism (Morris *et al.*, 2004).

Another fundamental question aside from how DNA is targeted by Dnmt for methylation modification is how methylated DNA leads to silencing of gene expression. So far, four models have become apparent. In the first, DNA methylation can prevent the transcriptional activator binding to the target DNA sequence, leading to direct repression of transactivation (Watt and Molloy, 1988). In the second, the Dnmt protein itself might physically link to HDAC and histone methylase (HMT) proteins. For example, Dnmt1 is itself associated with HDAC activity *in vivo*. HDAC1 has been shown to bind Dnmt1 and can purify methyltransferase activity from nuclear extracts. The transcriptional repression domain of Dnmt1 can also recruit HDAC directly (Fuks *et al.*, 2000). Dnmt3b can also couple with other chromatin-associated enzymatic activities common to transcriptional repression such as histone methylation (Geiman *et al.*, 2004). These raise the possibility that Dnmts mediate nonenzymatic roles in transcriptional silencing which rely on chromatin modification by complexing with other transcriptional repressor proteins (Robertson *et al.*, 2000; Rountree *et al.*, 2000). Besides the role in epigenetic modification of cytosine, Dnmts also function to silence gene expression directly through transcriptional repression. The third repression model is that DNA methylation within the gene body exerts a repressive effect on transcriptional elongation. Even though transcriptional silencing is often associated with promoter methylation, a considerable number of methylated CpGs are found in gene bodies including both intronic and exonic regions. When a transgene, which is methylated exclusively in a region downstream of the promoter, is introduced into a specific genomic site, it yields a clear decrease in the transgene expression relative to the unmethylated control. At the same time, there is a reduction in RNA polymerase II occupancy and chromatin accessibility. It is, therefore, proposed that the dense intragenic DNA methylation initiates the formation of a chromatin structure that reduces the efficiency of transcription elongation (Lorincz *et al.*, 2004). However, this model may not apply universally since DNA methylation of the gene body was also found to directly correlate with transcription level (Hellman and Chess, 2007). The fourth model of DNA methylation-induced gene transcription silencing is that methyl-CpG-binding proteins directly recognize methylated DNA and recruit corepressor molecules to silence transcription and also modify surrounding chromatin (Nan *et al.*, 1997, 1998). The family of methyl-CpG-binding proteins includes six members named MBD1–MBD4, Kaiso, and methyl-CpG binding protein 2 (MeCP2) (Fan and Hutnick, 2005; Klose and Bird, 2006). They are characterized by the methyl-CpG-binding domain (MBD), the protein motif responsible for binding methylated CpG dinucleotides. MBD1 can also interact with histone 3 lysine 9 (H3K9) methyltransferases to modify the surrounding chromatin by histone modification. The absence of MBD2 remarkably reduced tumorigenesis, implying that MBD2 contributes to tumor formation (Sansom *et al.*, 2003). MBD2 also reportedly shows robust demethylation ability *in vitro* (Bhattacharya *et al.*, 1999), but attempts to repeat the experiment failed in other laboratories (Bird, 2002). Though MBD3 contains a well-conserved MBD domain, it does not directly bind to methylated DNA because of a crucial amino acid substitution. As for MBD4, it is best known for its role in DNA repair (Hendrich *et al.*, 1999). It binds preferentially to G:T mismatches resulting from methyl-CpG deamination. Specifically, it may function to minimize mutations at methyl-CpG in particular. A novel methyl-CpG-binding protein named Kaiso lacks the MBD, but it can still recognize methylated DNA through zinc-finger domains (Prokhortchouk *et al.*, 2001). MeCP2 interacts specifically with methylated DNA and mediates transcriptional repression. MeCP2 binds tightly to chromosomes in a methylation-dependent manner. It in turn associates with a corepressor complex containing the transcriptional repressor mSin3A and HDAC. The relieved transcriptional repression by the deacetylase inhibitor TSA indicates that histone deacetylation is essential to this repression mechanism. In this way, MeCP2 links two global mechanisms of gene regulation, DNA methylation, and histone deacetylation in this transcriptional repression machinery.

IV. Role of DNA Methylation in the CNS and Neuropsychiatric Disorders

Epigenetic mechanisms, which include histone modification and DNA methylation, are believed to adjust the chromatin remodeling and accessibility of gene promoters (Jaenisch and Bird, 2003) to regulate neural adaptive gene expression. The essential roles of histone modification have been well demonstrated (Tsankova *et al.*, 2007). Different histone modifications, in combination or alone, define a specific epigenetic mark (histone code) that leads to different gene expression scenarios (Colvis *et al.*, 2005). Moreover, increased histone acetylation by HDAC inhibitors can even induce recovery of learning and memory (Fischer *et al.*, 2007; Guan *et al.*, 2009). However, the function of DNA methylation in the CNS is still unclear. Not as dynamic as reversible histone tail modifications, DNA methylation is a more static epigenetic mark, making it a less prominent candidate for dynamic neural gene expression regulation.

However, the importance of DNA methylation in the brain has been demonstrated by its association with some neurological disorders such as Rett syndrome (Amir *et al.*, 1999) and ICF syndrome (Hansen *et al.*, 1999). Mutations of MeCP2 have been found in about 80% of Rett syndrome patients (Amir *et al.*, 1999). Rett syndrome is one of the most common mental retardation diseases in females. Patients appear normal for the first 6–18 months of life, but gradually lose speech and motor skills. However, patients can still survive into adulthood, which suggests no progressive neurodegeneration. In the nervous system, MeCP2 is mainly expressed in neurons and shows an upregulation postnatally

(Jung *et al.*, 2003; Shahbazian *et al.*, 2002). Meanwhile, expression of MeCP2 is ubiquitous in rodents and humans, although protein levels vary between tissues. Mutations of MeCP2 only cause neurological defects, suggesting that there is a critical role for methyl-binding proteins/DNA methylation in the nervous system. By using transgenic mouse models (Chen *et al.*, 2001; Guy *et al.*, 2001), brain-specific deletion of MeCP2 at E12.5 resulted in an identical phenotype to that of the null mutation. Moreover, deletion of MeCP2 in only postnatal CNS neurons also led to a similar neuronal phenotype. These findings implicate that Rett syndrome is indeed caused by MeCP2 deficiency in the CNS as opposed to peripheral tissues, and the function of MeCP2 is more important in nervous system, especially in mature neurons (Kriaucionis and Bird, 2003). In addition, mutations of Dnmt3b gene have also been noticed in around 60% of ICF syndrome patients (Hansen *et al.*, 1999). Although the patients are characterized with immune defects, chromosomal instability as well as facial abnormalities, neurological defects including mental retardation is also common. Indeed, aberrant expression of genes regulating development and neurogenesis, which are relevant to the phenotypes, has been found in cell lines derived from ICF individuals (Jin *et al.*, 2008). Another case showing the importance of DNA methylation and/or methyl-CpG-binding proteins in brain function comes from MBD1 mutant mice. Besides reduced neuronal differentiation, adult MBD1 mutants have impaired spatial learning and significant reduction in long-term potentiation (LTP) of the hippocampus (Zhao *et al.*, 2003), potentially indicating the important role of DNA methylation in higher neural functioning. However, the search for methyl-CpG-binding protein regulated genes is still elusive (Tudor *et al.*, 2002). For instance, microarray study of MeCP2 null mice reveals only subtle transcriptional changes in the brain which could be explained as the sensitivity limitation of accurate detection of low-abundance transcripts or regional transcription differences. Alternatively, there may be some other DNA methylation or transcription-independent roles, which still require further investigation.

The studies of Dnmts can no doubt advance our understanding of the role of DNA methylation in the CNS. The expression of both the maintenance DNA methylation catalyzing enzyme Dnmt1 and the *de novo* DNA methyltransferases Dnmt3a and Dnmt3b have been characterized in the nervous system (Feng *et al.*, 2005; Goto *et al.*, 1994; Inano *et al.*, 2000; MacDonald *et al.*, 2005; Watanabe *et al.*, 2006). Dnmt1 is expressed at remarkably high levels in the embryonic nervous system. Additionally, almost all mature neurons in brains of adult mice express Dnmt1 at substantially high levels as compared with other organs. In the mouse brain, Dnmt3b is mainly expressed in early embryonic stages and neural progenitor cells before E15.5, whereas Dnmt3a is predominantly expressed in later embryonic stages through adulthood within neural precursor cells, maturing neurons, oligodendrocytes, and a subset of astrocytes (Feng *et al.*, 2005). As we discussed earlier, during cell replication, the newly synthesized DNA is hemimethylated (methylated parent strand and unmethylated daughter strand) and has to be precisely methylated on the daughter strand to maintain the mitotic inheritance of the methylation pattern from the parent strand. The maintenance DNA methyltransferase Dnmt1 is located at the replication fork and methylates hemimethylated DNA directly. Meanwhile, during germ cell development and embryogenesis, some DNA strands will switch from the unmethylated to the methylated state. *De novo* DNA methyltransferases Dnmt3a and Dnmt3b catalyze this process by adding methyl group to the unmethylated DNA to establish a new methylation pattern. Both *de novo* methylation and maintenance methylation happen during DNA replication. Generally, the DNA methylation pattern will not change after cell division, and afterward the Dnmt expression level in somatic cells diminishes greatly. That nondividing neurons still hold substantial levels of Dnmt1 and Dnmt3a (but not Dnmt3b) postnatally is highly indicative of their functional importance in the nervous system.

Interestingly, it was found that in schizophrenic postmortem brains, Dnmt1 is selectively overexpressed in the GABAergic neurons together with a downregulation of reelin and glutamic acid decarboxylase (GAD) 67 (Veldic *et al.*, 2004). Similarly, decreased expression of the $GABA_A$ receptor $\alpha 1$ gene was hypermethylated at its promoter region in suicidal/major depression disorder brains (Poulter *et al.*, 2008). Although the causal relationship is hard to establish, Dnmt1 knockdown in mouse primary neuronal culture was accompanied by increased reelin mRNA expression (Noh *et al.*, 2005). These data support the hypothesis that the aberrant gene expression changes in neuropsychiatric disorder brains may be the consequence of a Dnmt-mediated hypermethylation of the corresponding promoters.

Recent Dnmt inhibitor studies demonstrated that hippocampal LTP can be blocked by Dnmt inhibitors within rodent hippocampal slices (Levenson *et al.*, 2006; Nelson *et al.*, 2008). Moreover, these Dnmt inhibitors could also prevent rat memory formation following contextual fear conditioning (Miller and Sweatt, 2007). These support the notion that DNA methylation may target-specific genes involved in synaptic plasticity, learning, and memory. However, the Dnmt inhibitors (5-aza 2′-deoxyazacytidine and Zebularine) are nucleoside analogues that must be incorporated into DNA during DNA synthesis before they become inhibitors of Dnmts (Constantinides *et al.*, 1978; Jones *et al.*, 1982; Lyko and Brown, 2005). How they functioned in nondividing neurons (Tsankova *et al.*, 2007) or whether they actually worked in dividing cells such as glial cells and neural blast cells are unknown. The cytotoxicity effect of Dnmt inhibitors is also a concern for a broad application (Juttermann *et al.*, 1994; Lyko and Brown, 2005). Meanwhile, deficiency of selective Dnmt inhibitors makes it difficult to differentiate individual effects among different Dnmts. The future availability of less toxic, subtype specific Dnmt inhibitors will definitely improve our understanding of Dnmt/DNA methylation's role in the CNS. Alternatively, Dnmt mutant research can provide more direct evidence of DNA methylation's role in the nervous system.

Targeted mutation of Dnmt1 or Dnmt3b results in embryonic lethality before midgestation or E15.5, respectively (Li *et al.*, 1992; Okano *et al.*, 1999). A conventional deletion of Dnmt3a also causes lethality between postnatal weeks 2–3 (Okano *et al.*, 1999). Nervous system-specific conditional Dnmt knockouts were generated to better study the function of DNA methylation in the nervous system (Fan *et al.*, 2001; Hutnick *et al.*, 2009; Nguyen *et al.*, 2007). By utilizing the Cre/loxP system to inactivate Dnmt1 in neuroblasts of developing embryos (Fan *et al.*, 2001), it was found that the Dnmt1 deficiency in mitotic neural precursor cells resulted in DNA hypomethylation in daughter cells. Mice showing a nearly complete loss of Dnmt1 in the brain die immediately after birth because of respiratory distress. Mice whose brains have a low fraction of hypomethylated cells are viable but show a rapid loss of these cells and complete elimination within 3 weeks postnatally. This clearly proves that DNA methylation is required for neural cell survival. Interestingly, a similar nervous system-specific Dnmt3a knockout leads to fewer motor neurons in the hypoglossal nucleus (Nguyen *et al.*, 2007). Mice are born healthy, but degenerate and die prematurely. Their neuromuscular function and motor coordination defects display the role of Dnmt3a in neuromuscular control of motor movement (Nguyen *et al.*, 2007). Consistent with the predication that normal DNA methylation is required for the development of synaptic plasticity, studies of Emx1Cre triggered Dnmt1 deletion exclusively in telencephalic precursors found that thalamocortical LTP could not be induced. Dnmt1 deletion also blocked the development of somatosensory projection in the sensory cortex and induced striking cortical and hippocampal degeneration (Golshani *et al.*, 2005; Hutnick *et al.*, 2009) which suggest DNA methylation's roles in neuronal maturation.

DNA methylation also appears to play a role in neural cell fate specification. During development, neurons are always generated before astrocytes in the developing brain. How glial genes are restricted from developing neurons has long been of interest. There is increasing evidence that the transduction of these signals is regulated by epigenetic mechanisms at the level of DNA methylation and histone modification (Feng *et al.*, 2007; Hsieh and Gage, 2004, 2005). It has been shown that promoter demethylation of the glial fibrillary acidic protein (GFAP) gene, a gene normally expressed in astrocytes, is correlated with the GFAP gene transcription during astrogliogenesis (Takizawa *et al.*, 2001; Teter *et al.*, 1994). Conditional deletion of Dnmt1 in neural progenitor cells resulted in DNA hypomethylation and precocious astroglial differentiation (Fan *et al.*, 2005). Moreover, demethylation of genes in the gliogenic JAK–STAT pathway leads to enhanced activation of STATs, which triggers astrocyte differentiation. This finding provides a mechanism by which DNA methylation controls astrogliogenesis through STAT activation and promoter activation of GFAP glial marker genes.

However, all these findings are derived from Dnmt conditional knockouts in neural progenitor cells, so it is hard to discern whether the observed phenotypes are due to a loss of Dnmt in dividing progenitor cells or postmitotic neurons. A preliminary experiment indicated that at least the majority of the phenotypes are attributed to Dnmt's deletions from progenitor cells. In another strain of CamKIICre Dnmt1 conditional knockout mice, in which Dnmt1 deletion only occurs in postmitotic neurons, neither genomic hypomethylation nor obvious cell death were detected (Fan *et al.*, 2001). This may not be surprising because normal methylation patterns were in place before Dnmt1 was removed. However, the question remains why these neurons keep so much Dnmt postmitotically. It still cannot be ruled out that the Dnmt may still have some DNA methylation-independent or complementary (between Dnmt1 and Dnmt3a) functions in adult neurons. More than a decade ago, it was already proposed that Dnmt may serve to maintain DNA methylation after base-excision repair of the G:T mismatch that occurs spontaneously upon deamination of the methylated cytosine (Brooks *et al.*, 1996). Meanwhile, Dnmts may also be required to keep the DNA methylation pattern if methylation turnover (DNA demethylation) exists in adult neurons. Maintaining the genetic and epigenetic integrity in neurons could be critical since, unlike other cell types, nonrenewable postmitotic neurons have to function properly for their entire lifespan.

Active DNA methylation turnover in postmitotic neurons provides a compelling evidence for the existence of DNA demethylase and provides another function for Dnmts in neuronal cells. In contrast to blocking methylation of newly replicated DNA to passively demethylate DNA, active demethylation is replication independent. Some factors must take action to remove the methylation from already methylated DNA without cell division. In this sense, postmitotic neurons provide a good model to study with. Though Dnmts were recognized a long time ago, evidence for an enzyme exhibiting DNA demethylation capability in vertebrates is still lacking. By far, the most definitive DNA demethylases come only from plants (Zhu, 2008) instead of mammals. Though a number of mammalian DNA demethylases have been reported (for instance: MBD2 (Bhattacharya *et al.*, 1999), Gadd45a (Barreto *et al.*, 2007), Gadd45b (Ma *et al.*, 2009)), the results either could not be reproduced by other labs or led to controversial findings (Ooi and Bestor, 2008). One explanation of this difficulty is that intergenerational transmission of DNA methylation patterns within plants and mammals are fundamentally different (Ooi and Bestor, 2008). Plants transmit DNA methylation pattern with accretion of additional methylation during each generation, which makes the DNA demethylase necessary to prevent excess DNA methylation interference of nearby gene expression. Whereas mammalian germ cells may not require active DNA demethylation since they go through DNA methylation erasure first, followed by a reestablishment of the methylation pattern. Another possibility to explain the difficulty of identifying DNA demethylase is that DNA demethylation may actually be a coupling effect of several factors, as was recently noticed in the zebrafish, in which deaminase, glycosylase as well as Gadd45 work together to modulate DNA demethylation (Rai *et al.*, 2008).

Recently, there has been accumulating evidence proposes active demethylation in the CNS. For example, research of rodents' nurturing effect on offspring's stress response proved that DNA methylation in CNS could be quite dynamic (Francis *et al.*, 1999; Weaver *et al.*, 2004). As we know, early experiences can profoundly affect adult behavioral pattern and parenting style can influence a child's future personality and behavior. Interestingly, researchers found that offspring of high licking/grooming and arched-back nursing (LG-ABN) rat mothers are less fearful and show a more modest hypothalamic–pituitary–adrenal stress response in their adult lives (Liu *et al.*, 1997). How the maintenance of early environmental factors affects the phenotype is interesting. In a cross-fostering experiment, the biological offspring of a "low-LG-ABN" mother were raised by a "high-LG-ABN" mother. They showed a similar stress response to the normal offspring of a "high-LG-ABN" mother (Francis *et al.*, 1999). This implies that some nongenomic modulation may be responsible for this behavioral effect. It has been shown that there is a dynamic change of the epigenome around the glucocorticoid receptor (GR) gene promoter region, which leads to increased GR expression in adult offspring of "high-LG-ABN" mothers (Weaver *et al.*, 2004). A key transcription factor binding site, within the first exon of GR, is switched from the unmethylated state to the methylated state right after birth. Strikingly, exposure to "high-LG-ABN" mothering appears to demethylate this CpG site. Also histones surrounding the GR promoter became more acetylated. As a result, DNA methylation, together with histone acetylation, provides open access for GR transcriptional activation. The sustained GR expression is believed to maintain a long-term modest stress response. Additionally, this epigenetic regulation of GR expression could still be dynamic in the brain (Weaver *et al.*, 2005). A central administration of HDAC inhibitor in the adult rat reverses the stress response under "low-LG-ABN" care, whereas providing a methyl group donor for DNA methylation converts the stress response in "high-LG-ABN" offspring. This further proves the causal relationship between DNA methylation and long-term behavioral change. However, it is worthy noting that the patchy DNA demethylation evidence (Miller and Sweatt, 2007; Weaver *et al.*, 2004) identified so far in the CNS are still short of a cellular resolution. We are not aware whether they are from dividing neural progenitors and/or glia which is resulted from passive demethylation by blocking Dnmt access during DNA replication or from postmitotic neurons as a result of active demethylation by potential DNA demethylase.

V. Conclusion Remarks

DNA methylation provides an epigenetic means of neural gene expression regulation. In the past decade or so, the pivotal roles of DNA methylation as well as Dnmts in the CNS have been recognized in neural differentiation, cell survival, cell maturation, neural plasticity, as well as some neuropsychiatric disorders. However, a definitive function of Dnmt in the postmitotic neurons is still elusive. An active DNA methylation turnover in the neuronal cells also needs to be confirmed. In the future, genome-wide studies of both Dnmt targeting genes and neuronal DNA methylation change under both physiological and pathological conditions will further improve our understanding of the role of DNA methylation in neural function and potentially benefit neurological disorder therapies.

References

Amir, R. E., Van den Veyver, I. B., Wan, M., Tran, C. Q., Francke, U., and Zoghbi, H. Y. (1999). Rett syndrome is caused by mutations in X-linked MECP2, encoding methyl-CpG-binding protein 2. *Nat. Genet.* **23,** 185–188.

Barreto, G., Schafer, A., Marhold, J., Stach, D., Swaminathan, S. K., Handa, V., Doderlein, G., Maltry, N., Wu, W., Lyko, F., and Niehrs, C. (2007). Gadd45a promotes epigenetic gene activation by repair-mediated DNA demethylation. *Nature* **445,** 671–675.

Bestor, T., Laudano, A., Mattaliano, R., and Ingram, V. (1988). Cloning and sequencing of a cDNA encoding DNA methyltransferase of mouse cells. The carboxyl-terminal domain of the mammalian enzymes is related to bacterial restriction methyltransferases. *J. Mol. Biol.* **203,** 971–983.

Bhattacharya, S. K., Ramchandani, S., Cervoni, N., and Szyf, M. (1999). A mammalian protein with specific demethylase activity for mCpG DNA. *Nature* **397,** 579–583.

Bird, A. (2002). DNA methylation patterns and epigenetic memory. *Genes Dev.* **16,** 6–21.

Bird, A. (2007). Perceptions of epigenetics. *Nature* **447,** 396–398.

Bourc'his, D., Xu, G. L., Lin, C. S., Bollman, B., and Bestor, T. H. (2001). Dnmt3L and the establishment of maternal genomic imprints. *Science* **294,** 2536–2539.

Brenner, C., Deplus, R., Didelot, C., Loriot, A., Vire, E., De Smet, C., Gutierrez, A., Danovi, D., Bernard, D., Boon, T., Pelicci, P. G., Amati, B., *et al.* (2005). Myc represses transcription through recruitment of DNA methyltransferase corepressor. *EMBO J.* **24,** 336–346.

Brooks, P. J., Marietta, C., and Goldman, D. (1996). DNA mismatch repair and DNA methylation in adult brain neurons. *J. Neurosci.* **16,** 939–945.

Cameron, E. E., Bachman, K. E., Myohanen, S., Herman, J. G., and Baylin, S. B. (1999). Synergy of demethylation and histone deacetylase inhibition in the re-expression of genes silenced in cancer. *Nat. Genet.* **21,** 103–107.

Chen, R. Z., Akbarian, S., Tudor, M., and Jaenisch, R. (2001). Deficiency of methyl-CpG binding protein-2 in CNS neurons results in a Rett-like phenotype in mice. *Nat. Genet.* **27,** 327–331.

Colvis, C. M., Pollock, J. D., Goodman, R. H., Impey, S., Dunn, J., Mandel, G., Champagne, F. A., Mayford, M., Korzus, E., Kumar, A., Renthal, W., Theobald, D. E., *et al.* (2005). Epigenetic mechanisms and gene networks in the nervous system. *J. Neurosci.* **25,** 10379–10389.

Constantinides, P. G., Taylor, S. M., and Jones, P. A. (1978). Phenotypic conversion of cultured mouse embryo cells by aza pyrimidine nucleosides. *Dev. Biol.* **66,** 57–71.

Di Croce, L., Raker, V. A., Corsaro, M., Fazi, F., Fanelli, M., Faretta, M., Fuks, F., Lo Coco, F., Kouzarides, T., Nervi, C., Minucci, S., and Pelicci, P. G. (2002). Methyltransferase recruitment and DNA hypermethylation of target promoters by an oncogenic transcription factor. *Science* **295,** 1079–1082.

Fan, G., and Hutnick, L. (2005). Methyl-CpG binding proteins in the nervous system. *Cell Res.* **15,** 255–261.

Fan, G., Beard, C., Chen, R. Z., Csankovszki, G., Sun, Y., Siniaia, M., Biniszkiewicz, D., Bates, B., Lee, P. P., Kuhn, R., Trumpp, A., Poon, C., *et al.* (2001). DNA hypomethylation perturbs the function and survival of CNS neurons in postnatal animals. *J. Neurosci.* **21,** 788–797.

Fan, G., Martinowich, K., Chin, M. H., He, F., Fouse, S. D., Hutnick, L., Hattori, D., Ge, W., Shen, Y., Wu, H., ten Hoeve, J., Shuai, K., *et al.* (2005). DNA methylation controls the timing of astrogliogenesis through regulation of JAK-STAT signaling. *Development* **132,** 3345–3356.

Feng, J., Chang, H., Li, E., and Fan, G. (2005). Dynamic expression of *de novo* DNA methyltransferases Dnmt3a and Dnmt3b in the central nervous system. *J. Neurosci. Res.* **79,** 734–746.

Feng, J., Fouse, S., and Fan, G. (2007). Epigenetic regulation of neural gene expression and neuronal function. *Pediatr. Res.* **61,** 58R–63R.

Fischer, A., Sananbenesi, F., Wang, X., Dobbin, M., and Tsai, L. H. (2007). Recovery of learning and memory is associated with chromatin remodelling. *Nature* **447,** 178–182.

Francis, D., Diorio, J., Liu, D., and Meaney, M. J. (1999). Nongenomic transmission across generations of maternal behavior and stress responses in the rat. *Science* **286,** 1155–1158.

Fuks, F., Burgers, W. A., Brehm, A., Hughes-Davies, L., and Kouzarides, T. (2000). DNA methyltransferase Dnmt1 associates with histone deacetylase activity. *Nat. Genet.* **24,** 88–91.

Ge, Y. Z., Pu, M. T., Gowher, H., Wu, H. P., Ding, J. P., Jeltsch, A., and Xu, G. L. (2004). Chromatin targeting of *de novo* DNA methyltransferases by the PWWP domain. *J. Biol. Chem.* **279,** 25447–25454.

Geiman, T. M., Sankpal, U. T., Robertson, A. K., Zhao, Y., Zhao, Y., and Robertson, K. D. (2004). DNMT3B interacts with hSNF2H chromatin remodeling enzyme, HDACs 1 and 2, and components of the histone methylation system. *Biochem. Biophys. Res. Commun.* **318,** 544–555.

Goll, M. G., and Bestor, T. H. (2005). Eukaryotic cytosine methyltransferases. *Annu. Rev. Biochem.* **74,** 481–514.

Golshani, P., Hutnick, L., Schweizer, F., and Fan, G. (2005). Conditional Dnmt1 deletion in dorsal forebrain disrupts development of somatosensory barrel cortex and thalamocortical long-term potentiation. *Thalamus Relat. Syst.* **3,** 227–233.

Goto, K., Numata, M., Komura, J. I., Ono, T., Bestor, T. H., and Kondo, H. (1994). Expression of DNA methyltransferase gene in mature and immature neurons as well as proliferating cells in mice. *Differentiation* **56,** 39–44.

Guan, Z., Giustetto, M., Lomvardas, S., Kim, J. H., Miniaci, M. C., Schwartz, J. H., Thanos, D., and Kandel, E. R. (2002). Integration of long-term-memory-related synaptic plasticity involves bidirectional regulation of gene expression and chromatin structure. *Cell* **111,** 483–493.

Guan, J. S., Haggarty, S. J., Giacometti, E., Dannenberg, J. H., Joseph, N., Gao, J., Nieland, T. J., Zhou, Y., Wang, X., Mazitschek, R., Bradner, J. E., DePinho, R. A., *et al.* (2009). HDAC2 negatively regulates memory formation and synaptic plasticity. *Nature* **459,** 55–60.

Guy, J., Hendrich, B., Holmes, M., Martin, J. E., and Bird, A. (2001). A mouse Mecp2-null mutation causes neurological symptoms that mimic Rett syndrome. *Nat. Genet.* **27,** 322–326.

Hansen, R. S., Wijmenga, C., Luo, P., Stanek, A. M., Canfield, T. K., Weemaes, C. M., and Gartler, S. M. (1999). The DNMT3B DNA methyltransferase gene is mutated in the ICF immunodeficiency syndrome. *Proc. Natl. Acad. Sci. USA* **96,** 14412–14417.

Hellman, A., and Chess, A. (2007). Gene body-specific methylation on the active X chromosome. *Science* **315,** 1141–1143.

Hendrich, B., Hardeland, U., Ng, H. H., Jiricny, J., and Bird, A. (1999). The thymine glycosylase MBD4 can bind to the product of deamination at methylated CpG sites. *Nature* **401,** 301–304.

Hermann, A., Goyal, R., and Jeltsch, A. (2004). The Dnmt1 DNA-(cytosine-C5)-methyltransferase methylates DNA processively with high preference for hemimethylated target sites. *J. Biol. Chem.* **279,** 48350–48359.

Holliday, R. (1999). Is there an epigenetic component in long-term memory? *J. Theor. Biol.* **200,** 339–341.

Hsieh, C. L. (1999). *In vivo* activity of murine *de novo* methyltransferases, Dnmt3a and Dnmt3b. *Mol. Cell. Biol.* **19,** 8211–8218.

Hsieh, J., and Gage, F. H. (2004). Epigenetic control of neural stem cell fate. *Curr. Opin. Genet. Dev.* **14,** 461–469.

Hsieh, J., and Gage, F. H. (2005). Chromatin remodeling in neural development and plasticity. *Curr. Opin. Cell Biol.* **17,** 664–671.

Hutnick, L. K., Golshani, P., Namihira, M., Xue, Z., Matynia, A., Yang, X. W., Silva, A. J., Schweizer, F. E., and Fan, G. (2009). DNA hypomethylation restricted to the murine forebrain induces cortical degeneration and impairs postnatal neuronal maturation. *Hum. Mol. Genet.* **18,** 2875–2888.

Inano, K., Suetake, I., Ueda, T., Miyake, Y., Nakamura, M., Okada, M., and Tajima, S. (2000). Maintenance-type DNA methyltransferase is highly expressed in post-mitotic neurons and localized in the cytoplasmic compartment. *J. Biochem.* **128,** 315–321.

Jaenisch, R., and Bird, A. (2003). Epigenetic regulation of gene expression: How the genome integrates intrinsic and environmental signals. *Nat. Genet.* **33**(Suppl.), 245–254.

Jakobsson, J., Cordero, M. I., Bisaz, R., Groner, A. C., Busskamp, V., Bensadoun, J. C., Cammas, F., Losson, R., Mansuy, I. M., Sandi, C., and Trono, D. (2008). KAP1-mediated epigenetic repression in the forebrain modulates behavioral vulnerability to stress. *Neuron* **60,** 818–831.

Jenuwein, T., and Allis, C. D. (2001). Translating the histone code. *Science* **293,** 1074–1080.

Jin, B., Tao, Q., Peng, J., Soo, H. M., Wu, W., Ying, J., Fields, C. R., Delmas, A. L., Liu, X., Qiu, J., and Robertson, K. D. (2008). DNA methyltransferase 3B (DNMT3B) mutations in ICF syndrome lead to altered epigenetic modifications and aberrant expression of genes regulating development, neurogenesis and immune function. *Hum. Mol. Genet.* **17,** 690–709.

Jones, P. A., Taylor, S. M., Mohandas, T., and Shapiro, L. J. (1982). Cell cycle-specific reactivation of an inactive X-chromosome locus by 5-azadeoxycytidine. *Proc. Natl. Acad. Sci. USA* **79,** 1215–1219.

Jung, B. P., Jugloff, D. G., Zhang, G., Logan, R., Brown, S., and Eubanks, J. H. (2003). The expression of methyl CpG binding factor MeCP2 correlates with cellular differentiation in the developing rat brain and in cultured cells. *J. Neurobiol.* **55,** 86–96.

Juttermann, R., Li, E., and Jaenisch, R. (1994). Toxicity of 5-aza-2′-deoxycytidine to mammalian cells is mediated primarily by covalent trapping of DNA methyltransferase rather than DNA demethylation. *Proc. Natl. Acad. Sci. USA* **91,** 11797–11801.

Klose, R. J., and Bird, A. P. (2006). Genomic DNA methylation: The mark and its mediators. *Trends Biochem. Sci.* **31,** 89–97.

Klose, R. J., Kallin, E. M., and Zhang, Y. (2006). JmjC-domain-containing proteins and histone demethylation. *Nat. Rev. Genet.* **7,** 715–727.

Korzus, E., Rosenfeld, M. G., and Mayford, M. (2004). CBP histone acetyltransferase activity is a critical component of memory consolidation. *Neuron* **42,** 961–972.

Kouzarides, T. (2007). Chromatin modifications and their function. *Cell* **128,** 693–705.
Kriaucionis, S., and Bird, A. (2003). DNA methylation and Rett syndrome. *Hum. Mol. Genet.* **12**(Spec. No. 2), R221–R227.
Levenson, J. M., Roth, T. L., Lubin, F. D., Miller, C. A., Huang, I. C., Desai, P., Malone, L. M., and Sweatt, J. D. (2006). Evidence that DNA (cytosine-5) methyltransferase regulates synaptic plasticity in the hippocampus. *J. Biol. Chem.* **281,** 15763–15773.
Li, E., Bestor, T. H., and Jaenisch, R. (1992). Targeted mutation of the DNA methyltransferase gene results in embryonic lethality. *Cell* **69,** 915–926.
Liu, D., Diorio, J., Tannenbaum, B., Caldji, C., Francis, D., Freedman, A., Sharma, S., Pearson, D., Plotsky, P. M., and Meaney, M. J. (1997). Maternal care, hippocampal glucocorticoid receptors, and hypothalamic-pituitary-adrenal responses to stress. *Science* **277,** 1659–1662.
Lorincz, M. C., Dickerson, D. R., Schmitt, M., and Groudine, M. (2004). Intragenic DNA methylation alters chromatin structure and elongation efficiency in mammalian cells. *Nat. Struct. Mol. Biol.* **11,** 1068–1075.
Lyko, F., and Brown, R. (2005). DNA methyltransferase inhibitors and the development of epigenetic cancer therapies. *J. Natl. Cancer Inst.* **97,** 1498–1506.
Ma, D. K., Jang, M. H., Guo, J. U., Kitabatake, Y., Chang, M. L., Pow-Anpongkul, N., Flavell, R. A., Lu, B., Ming, G. L., and Song, H. (2009). Neuronal activity-induced Gadd45b promotes epigenetic DNA demethylation and adult neurogenesis. *Science* **323,** 1074–1077.
MacDonald, J. L., Gin, C. S., and Roskams, A. J. (2005). Stage-specific induction of DNA methyltransferases in olfactory receptor neuron development. *Dev. Biol.* **288,** 461–473.
Miller, C. A., and Sweatt, J. D. (2007). Covalent modification of DNA regulates memory formation. *Neuron* **53,** 857–869.
Morris, K. V., Chan, S. W., Jacobsen, S. E., and Looney, D. J. (2004). Small interfering RNA-induced transcriptional gene silencing in human cells. *Science* **305,** 1289–1292.
Nan, X., Campoy, F. J., and Bird, A. (1997). MeCP2 is a transcriptional repressor with abundant binding sites in genomic chromatin. *Cell* **88,** 471–481.
Nan, X., Ng, H. H., Johnson, C. A., Laherty, C. D., Turner, B. M., Eisenman, R. N., and Bird, A. (1998). Transcriptional repression by the methyl-CpG-binding protein MeCP2 involves a histone deacetylase complex. *Nature* **393,** 386–389.
Nelson, E. D., Kavalali, E. T., and Monteggia, L. M. (2008). Activity-dependent suppression of miniature neurotransmission through the regulation of DNA methylation. *J. Neurosci.* **28,** 395–406.
Nguyen, S., Meletis, K., Fu, D., Jhaveri, S., and Jaenisch, R. (2007). Ablation of *de novo* DNA methyltransferase Dnmt3a in the nervous system leads to neuromuscular defects and shortened lifespan. *Dev. Dyn.* **236,** 1663–1676.
Noh, J. S., Sharma, R. P., Veldic, M., Salvacion, A. A., Jia, X., Chen, Y., Costa, E., Guidotti, A., and Grayson, D. R. (2005). DNA methyltransferase 1 regulates reelin mRNA expression in mouse primary cortical cultures. *Proc. Natl. Acad. Sci. USA* **102,** 1749–1754.
Okano, M., Xie, S., and Li, E. (1998). Dnmt2 is not required for *de novo* and maintenance methylation of viral DNA in embryonic stem cells. *Nucleic Acids Res.* **26,** 2536–2540.
Okano, M., Bell, D. W., Haber, D. A., and Li, E. (1999). DNA methyltransferases Dnmt3a and Dnmt3b are essential for *de novo* methylation and mammalian development. *Cell* **99,** 247–257.
Ooi, S. K., and Bestor, T. H. (2008). The colorful history of active DNA demethylation. *Cell* **133,** 1145–1148.
Petrij, F., Giles, R. H., Dauwerse, H. G., Saris, J. J., Hennekam, R. C., Masuno, M., Tommerup, N., van Ommen, G. J., Goodman, R. H., Peters, D. J., *et al.* (1995). Rubinstein-Taybi syndrome caused by mutations in the transcriptional co-activator CBP. *Nature* **376,** 348–351.

Poulter, M. O., Du, L., Weaver, I. C., Palkovits, M., Faludi, G., Merali, Z., Szyf, M., and Anisman, H. (2008). GABAA receptor promoter hypermethylation in suicide brain: Implications for the involvement of epigenetic processes. *Biol. Psychiatry* **64,** 645–652.

Pradhan, S., and Esteve, P. O. (2003). Mammalian DNA (cytosine-5) methyltransferases and their expression. *Clin. Immunol.* **109,** 6–16.

Pradhan, S., Bacolla, A., Wells, R. D., and Roberts, R. J. (1999). Recombinant human DNA (cytosine-5) methyltransferase. I. Expression, purification, and comparison of *de novo* and maintenance methylation. *J. Biol. Chem.* **274,** 33002–33010.

Prokhortchouk, A., Hendrich, B., Jorgensen, H., Ruzov, A., Wilm, M., Georgiev, G., Bird, A., and Prokhortchouk, E. (2001). The p120 catenin partner Kaiso is a DNA methylation-dependent transcriptional repressor. *Genes Dev.* **15,** 1613–1618.

Rai, K., Huggins, I. J., James, S. R., Karpf, A. R., Jones, D. A., and Cairns, B. R. (2008). DNA demethylation in zebrafish involves the coupling of a deaminase, a glycosylase, and gadd45. *Cell* **135,** 1201–1212.

Robertson, K. D., Ait-Si-Ali, S., Yokochi, T., Wade, P. A., Jones, P. L., and Wolffe, A. P. (2000). DNMT1 forms a complex with Rb, E2F1 and HDAC1 and represses transcription from E2F-responsive promoters. *Nat. Genet.* **25,** 338–342.

Rountree, M. R., Bachman, K. E., and Baylin, S. B. (2000). DNMT1 binds HDAC2 and a new co-repressor, DMAP1, to form a complex at replication foci. *Nat. Genet.* **25,** 269–277.

Sansom, O. J., Berger, J., Bishop, S. M., Hendrich, B., Bird, A., and Clarke, A. R. (2003). Deficiency of Mbd2 suppresses intestinal tumorigenesis. *Nat. Genet.* **34,** 145–147.

Shahbazian, M. D., Antalffy, B., Armstrong, D. L., and Zoghbi, H. Y. (2002). Insight into Rett syndrome: MeCP2 levels display tissue- and cell-specific differences and correlate with neuronal maturation. *Hum. Mol. Genet.* **11,** 115–124.

Shi, Y., Lan, F., Matson, C., Mulligan, P., Whetstine, J. R., Cole, P. A., Casero, R. A., and Shi, Y. (2004). Histone demethylation mediated by the nuclear amine oxidase homolog LSD1. *Cell* **119,** 941–953.

Suetake, I., Shinozaki, F., Miyagawa, J., Takeshima, H., and Tajima, S. (2004). DNMT3L stimulates the DNA methylation activity of Dnmt3a and Dnmt3b through a direct interaction. *J. Biol. Chem.* **279,** 27816–27823.

Takizawa, T., Nakashima, K., Namihira, M., Ochiai, W., Uemura, A., Yanagisawa, M., Fujita, N., Nakao, M., and Taga, T. (2001). DNA methylation is a critical cell-intrinsic determinant of astrocyte differentiation in the fetal brain. *Dev. Cell* **1,** 749–758.

Teter, B., Osterburg, H. H., Anderson, C. P., and Finch, C. E. (1994). Methylation of the rat glial fibrillary acidic protein gene shows tissue-specific domains. *J. Neurosci. Res.* **39,** 680–693.

Tsankova, N., Renthal, W., Kumar, A., and Nestler, E. J. (2007). Epigenetic regulation in psychiatric disorders. *Nat. Rev. Neurosci.* **8,** 355–367.

Tudor, M., Akbarian, S., Chen, R. Z., and Jaenisch, R. (2002). Transcriptional profiling of a mouse model for Rett syndrome reveals subtle transcriptional changes in the brain. *Proc. Natl. Acad. Sci. USA* **99,** 15536–15541.

Veldic, M., Caruncho, H. J., Liu, W. S., Davis, J., Satta, R., Grayson, D. R., Guidotti, A., and Costa, E. (2004). DNA-methyltransferase 1 mRNA is selectively overexpressed in telencephalic GABAergic interneurons of schizophrenia brains. *Proc. Natl. Acad. Sci. USA* **101,** 348–353.

Watanabe, D., Uchiyama, K., and Hanaoka, K. (2006). Transition of mouse *de novo* methyltransferases expression from Dnmt3b to Dnmt3a during neural progenitor cell development. *Neuroscience* **142,** 727–737.

Watt, F., and Molloy, P. L. (1988). Cytosine methylation prevents binding to DNA of a HeLa cell transcription factor required for optimal expression of the adenovirus major late promoter. *Genes Dev.* **2,** 1136–1143.

Weaver, I. C., Cervoni, N., Champagne, F. A., D'Alessio, A. C., Sharma, S., Seckl, J. R., Dymov, S., Szyf, M., and Meaney, M. J. (2004). Epigenetic programming by maternal behavior. *Nat. Neurosci.* **7,** 847–854.

Weaver, I. C., Champagne, F. A., Brown, S. E., Dymov, S., Sharma, S., Meaney, M. J., and Szyf, M. (2005). Reversal of maternal programming of stress responses in adult offspring through methyl supplementation: Altering epigenetic marking later in life. *J. Neurosci.* **25,** 11045–11054.

Xie, S., Wang, Z., Okano, M., Nogami, M., Li, Y., He, W. W., Okumura, K., and Li, E. (1999). Cloning, expression and chromosome locations of the human DNMT3 gene family. *Gene* **236,** 87–95.

Xu, G. L., Bestor, T. H., Bourc'his, D., Hsieh, C. L., Tommerup, N., Bugge, M., Hulten, M., Qu, X., Russo, J. J., and Viegas-Pequignot, E. (1999). Chromosome instability and immunodeficiency syndrome caused by mutations in a DNA methyltransferase gene. *Nature* **402,** 187–191.

Zhao, X., Ueba, T., Christie, B. R., Barkho, B., McConnell, M. J., Nakashima, K., Lein, E. S., Eadie, B. D., Willhoite, A. R., Muotri, A. R., Summers, R. G., Chun, J., *et al.* (2003). Mice lacking methyl-CpG binding protein 1 have deficits in adult neurogenesis and hippocampal function. *Proc. Natl. Acad. Sci. USA* **100,** 6777–6782.

Zhu, J. K. (2008). Epigenome sequencing comes of age. *Cell* **133,** 395–397.

HERITABILITY OF STRUCTURAL BRAIN TRAITS: AN ENDOPHENOTYPE APPROACH TO DECONSTRUCT SCHIZOPHRENIA

Nil Kaymaz*,† and J. van Os*,‡

*Department of Psychiatry and Psychology, University of Maastricht, P.O. Box 616 (DRT10) Maastricht, The Netherlands

†Mediant GGZ/Mental Health Care, Postbus 775, 7500 AT Enschede, The Netherlands

‡Division of Psychological Medicine, Institute of Psychiatry, London, United Kingdom

Structural brain phenotypes are quantitative traits showing considerable variation in human populations. Quantitative structural brain abnormalities are also repeatedly reported in patients with psychiatric disorders such as schizophrenia. Studying the genetic and environmental causes of these differences might therefore highlight biological mechanisms underlying neuroanatomical phenotypes and directly result in the identification of risk factors for schizophrenia. Heritability estimates indicate a strong genetic component contributing to neuroanatomical phenotypes. Brain structure volumes have substantial heritability

INTERNATIONAL REVIEW OF NEUROBIOLOGY, VOL. 89
DOI: 10.1016/S0074-7742(09)89005-3

rates ranging from high (70–95%) for total brain volume, cerebellar, gray and white matter, and corpus callosum, to moderate (40–70%) determined for the hippocampus, the four lobes (frontal, temporal, occipital, and parietal lobe), temporal horn volume, brain parenchyma, white matter hyperintensity, and planum temporal asymmetry. Middle structures of the brain show high heritability scores for the deeper structures (ontogenetically earlier formed) and moderate heritability scores for the surface structures. Structures formed earlier in development show consistently higher heritability rates than brain structures formed later in development, for example, surface structures, which seem to be influenced by environmental factors. Even higher heritability reaching 0.99 for total brain volume are estimated in nonhuman primate (NHP) models employing inbred extended pedigree and highly uniform rearing conditions, reducing the effects of environmental factors. Applying highly heritable structural brain phenotypes may serve as an endophenotype for gene mapping studies and lead to identification of genes that are involved in the regulation of human brain volume and the biological mechanisms involved in the causal mechanisms of psychiatric disorders.

I. Introduction

A. Endophenotypic Approach to Studying Schizophrenia

Schizophrenia is a highly heritable disorder (Cannon *et al.*, 1998; Cardno and Gottesman, 2000; Sullivan *et al.*, 2003) characterized by delusions, hallucinations, disorganized speech (frequent derailment or incoherence), grossly disorganized or catatonic behavior, and negative symptoms, that is, affective flattening, alogia (poverty of speech) and avolition (general lack of desire, interest or motivation to pursue meaningful goals) (DSM-IV-Tr, 2000). It is believed that this complex phenotype of schizophrenia is a result of underlying complex genetic architecture involving interactions between multiple loci and environmental factors (van Os *et al.*, 2008). Another source of difficulty is the reliability of phenotypic assessments. Schizophrenia diagnostics employ a patient's self-reports rather than biochemical, electrophysiological, or other reliably measurable biomarkers. The validity of schizophrenia as a diagnostic entity has therefore been criticized by a number of scientists and clinicians as lacking in scientific validity and diagnostic reliability (Bentall *et al.*, 2004; Boyle, 2002; Kendell and Jablensky, 2003; Linscott *et al.*, 2009; van Os, 2009; van Os and Murray, 2008). Due to the diagnostic heterogeneity and the genetic and phenotypic complexity of schizophrenia, as well as other complex brain-related disorders, identification of a genetic basis of this disease is a very challenging task.

The endophenotypic approach is one of several strategies that have been adopted to deal with the complexities of these disorders and identify underlying genes (Gottesman and Gould, 2003). The principle of this approach is to reduce complex phenotypes into components (neurophysiological, biochemical, endocrine, cognitive, neuropsychological, or neuroanatomical phenotypes) that could be reliably quantified and are under close regulation by genetic variation. The endophenotype (or intermediate phenotype) is therefore a kind of biomarker that, unlike disease diagnosis, has a clearer underlying biological and genetic basis. They are expected to be genetically correlated with disease liability, and can be measured in both affected and unaffected individuals. Since quantitative endophenotypes are generally closer to the action of the gene and are measurable in both unaffected and controls, they exhibit higher genetic signal-to-noise ratios and are anticipated to provide a greater power to localize and identify disease-related quantitative trait loci (QTLs) than affection status alone (Blangero *et al.*, 2003).

Human studies often use endophenotypes that are related to evolutionary conserved traits and therefore can be extended to animal models.

B. Neuroanatomic Measures as Endophenotype

Volumes of various brain structures are quantitative phenotypes that can be reliably measured on MRI scans using either automated (voxel-based morphometry method, VBM; deformation-based morphometry; surface-based morphometry) or manual tracing (region and volume of interest, ROI and VOI method) (Ashburner and Friston, 2000; Good *et al.*, 2001a; Thompson and Toga, 1997; Thompson *et al.*, 1996a,b). Brain structure measurements show considerable interindividual variation in general populations and differences between healthy and affected persons, and are therefore considered promising candidate endophenotypes that might facilitate investigations of a genetic basis into the natural variation of brain-related traits in populations and for psychiatric disorders such as schizophrenia. When measuring brain structure and the heritability of these structures, normal neuroanatomical variations in the human brain should be taken into account (Allen *et al.*, 2002). These neuroanatomical variations are under the influence of variables such as gender (Cosgrove *et al.*, 2007; Giedd, 2008) and age (Carne *et al.*, 2006; Good *et al.*, 2001b), or other components of genetic variety such as height (Kappelman, 1993). Intelligence seems to be related to brain volume as well. In humans, brain volume is a quantitative trait with high heritability (Posthuma *et al.*, 2002b; Thompson *et al.*, 2001, 2002). Previous studies have shown that brain volume is also correlated with general intelligence, working memory, perceptual organization, and processing speed, which are also highly heritable (Posthuma *et al.*, 2003; Roth and Dicke, 2005). In recent years, it has been discovered that the structure of the adult human brain changes when new cognitive or motor skills, including vocabulary, are learned (Lee *et al.*, 2007), and that this structural neuroplasticity (increased gray matter volume), after 3 months of training a visual motor skill, seems to last for at least 3 months without further practicing (Driemeyer *et al.*, 2008; Ilg *et al.*, 2008). Another example is the impact of handedness, which has a significant genetic component (Annett, 1974; McManus and Bryden, 1991) and is strongly related with cerebral asymmetry (Alexander and Annett, 1996; Geschwind and Galaburda, 1985). To what extend the cerebral asymmetry is also heritable remains unclear (Geschwind *et al.*, 2002). The neuroanatomical phenotype and the heritability of brain structures varies among individuals with a neuropsychiatric disorder, meaning that the influence the disease state has on the heritability is such that the morphology of brain structures can be altered by the disease state and these morphologically altered brain structures can be inherited. For example, there are studies showing small hippocampal volumes in adult humans with recurrent major depression (Bremner *et al.*, 2000; Sheline *et al.*, 1999) and posttraumatic stress disorder (Bremner *et al.*, 1995). Hippocampal morphology is also altered by stress in carefully controlled studies with rodents (Gould and Tanapat, 1999; Magarinos *et al.*, 1995; Sapolsky, 2000). These smaller hippocampi have been taken as evidence that stress-related disorders induce hippocampal volume loss (Sapolsky, 2000; Sheline *et al.*, 1999). One possible pathway by which altered brain morphology changes might be inherited is that small hippocampal volumes are inherited and act as predisposing factors toward the development of psychiatric disorders that are triggered by stress (Gurvits *et al.*, 1996; Sapolsky, 2000).

When measuring brain structure volumes and calculating heritability estimates, these variables, having a strong genetic component themselves, should be taken into account.

C. Heritability Estimates to Determine Contribution of Genetic Factors to Structural Brain Traits

(*to study genetics of both natural and disease-related variation*)

Twin and adoption studies have shown substantial genetic influences are involved in the risk of developing schizophrenia (Cannon *et al.*, 1998; Cardno *et al.*, 2002; Kendler and Diehl, 1993; McGuffin *et al.*, 1984). The identification of predisposing genes has been hampered by difficulties in detecting nonpenetrant carriers and by uncertainties concerning the extent of locus of heterogeneity (McDonald and Murphy, 2003). By studying the inheritance of endophenotypes, we can increase the power to detect the genes involved in clarifying the pathways leading from genetic predisposition to clinical disorder

(Gottesman and Gould, 2003). One of the most extensively studied endophenotypes in schizophrenia research is the study of structural brain abnormalities. These structural changes are well established in schizophrenia. The most robust findings from meta-analyses of these structural brain abnormalities is increased total ventricular volume, reduction of whole brain and intracranial volume, reduced hippocampal and amygdalar volume (Lawrie and Abukmeil, 1998; Ward *et al.*, 1996; Wright *et al.*, 2000). These brain abnormalities are also present in unaffected family members of patients with schizophrenia, to a less degree, but significantly more than in controls according to a meta-analysis reporting on brain changes in unaffected family members of patients with schizophrenia (Boos *et al.*, 2007; van Haren *et al.*, 2004).

One hypothesis in using these structural abnormalities as endophenotypes, in order to get closer to the action of the genes, suggests that there might be an overlap in the neurodevelopmental genes responsible for both the volume change and the development of the illness, that is, the existence of a genetic correlation. Schizophrenia may involve genetically determined pathological processes of early brain development which continue to unfold as the brain matures through neuronal loss and synaptic pruning during adolescence. Neurodevelopmental abnormalities then lead to the activation of pathological neural circuits, which respond to environmental stressors, leading to the emergence of symptoms (Deutch, 1993). This hypothesis is supported by the finding that MRI abnormalities are present at the onset of the illness, and progress very slowly if at all (Weinberger, 1995). A contrasting view on the etiology of schizophrenia is that the relationship between schizophrenia and brain volumes might be environmental in nature. For example, perinatal trauma has been shown to be an important determinant of some brain structure abnormalities in schizophrenia (Cannon *et al.*, 2003; McNeil, 2000; Verdoux *et al.*, 1997).

Defining to what degree the development of these brain structures is of genetic origin, that is, the heritability of brain structures, and using these brain structures as an endophenotype may be helpful in better identifying the action of these neurodevelopmental genes. Heritability estimates for brain structures in healthy persons and persons with schizophrenia will help to differentiate between disease related alterations in brain morphology and genetic influences (heritability).

Twin studies are considered important in investigating genetic influences on variation in human brain morphology in healthy individuals and those with schizophrenia. The twin model is particularly helpful in determining the relative contribution of genetic, common, and unique environmental influences on variation in brain structures (Posthuma and Boomsma, 2000). Heritability estimates of brain structure are usually based on data from monozygotic twins (MZ, who are nearly genetically identical) and dizygotic twins (DZ, who share on average 50% of their segregating genes). If heritability estimates are based on the assumption that for a certain brain measure, MZ twin pairs will resemble each other more closely than DZ twin pairs, it can be inferred that variation of the brain measure is heritable. However, in addition to genetic influences, common (or shared) environmental influences may play a role in explaining resemblances. The effects of shared environmental factors may be suggested when correlations in DZ twins are larger than half of the MZ correlation (Boomsma *et al.*, 2002). The effects of unique environmental factors are obtained from the extent to which MZ twins do not resemble each other. However, the twin method has been criticized for its nongeneralizability due to differences in intrauterine and family environment differences between the twins and compared with singletons (Price, 1950).

Heritability estimates and the search for genes in diseased and nondiseased humans are difficult due to the impact of environmental factors. Studies in nonhuman primates (NHPs) might be helpful in elucidating the impact of environmental factors.

D. Nonhuman Primate Model for Studying Brain Neurostructural Studies

Obstacles in investigating complex traits in human subjects involve the hard-to-control and extensive effects of environment, personal development, and medical history including administrated medications. To a large extent, these problems can be overcome by applying an animal model. Particularly useful are NHPs due to the high conservation of anatomical, neurophysiological and cognitive traits as well genetic sequence with humans. NHP pedigrees have helped studies of various complex traits including the neruoanatomical due mostly to the availability of pedigree facilitating genetic studies, established genetic relationships among pedigree animals, well-documented medical and developmental history and uniform rearing conditions reducing the influence of environmental factors.

To disentangle the possible mechanisms by which genetic and environmental factors contribute to the morphological abnormalities in the brain animal models may be used, specifically NHPs, for example, monkey brains, in which the influence of environment is reduced. The evolution of cognitive function in hominoids, for example, depends largely on our understanding of the organization of the frontal lobes in extant humans and apes (Semendeferi *et al.*, 1997). There are studies with findings showing similarities to human brains, such as the possibility of microstructural plasticity in the NHP hippocampus (Sapolsky *et al.*, 1990; Siegel *et al.*, 1993; Uno *et al.*, 1989). Other studies show the impact of environmental factors on the volumes of different brain structures, such as the affect of differential rearing on the corpus callosum size in rhesus monkeys (Sánchez *et al.*, 1998). Studies on brain characteristics and the genetic variation of regional brain morphologies of NHPs will help us understand the meaning of the brain characteristics in humans as well. They will also help us determine whether certain assumptions, such as those about certain brain characteristics being strictly unique to humans, are correct. For example, it is often claimed that the frontal lobe is disproportionately larger in humans than in other species, but conflicting reports exist on this issue. Results of a study by Semendeferi *et al.* (1997) indicate that although the absolute volume of the brain and the frontal lobe is the largest in humans, the relative size of the frontal lobe is similar across hominoids and humans do not have a larger frontal lobe than expected from a primate brain of human size. Other important functional parts of the brain, like area 10 in cortical areas of the frontal lobe, which is involved in higher cognitive functions, also seem to form the frontal pole in chimpanzee, bonobo, orangutan, macaque, and gibbon brains. Area 10 has similar cytoarchitectonic features in the hominoid brain, but aspects of its organization vary slightly across species, including relative width of its cortical layers and the space available for connections (Semendeferi *et al.*, 2001). Although some features of the human brain, like the asymmetric Broca's area and the planum temporal in Wernicke's posterior receptive language area in the left hemisphere, were thought to be unique to humans, studies in great apes (Cantalupo and Hopkins, 2001) and chimpanzees (Gannon *et al.*, 1998) show the same anatomic hemispheric asymmetry in the Broca's area and planum temporal, respectively.

II. Heritability of Brain Structure Phenotypes

Heritability estimates allow for determining which neuroanatomical measures are traits with marked genetic components and could therefore be interesting both from the perspective of natural variation in brain related processes in population and as an attractive endopehnotype for schizophrenia studies. Investigating heritability in various aspects in healthy individuals, in relation to schizophrenia and in the NHP model, may answer two essential questions: (1) to what degree are brain volumes genetically determined? and, (2) which brain volumes are mostly heritable and therefore useful as endophenotypes for genetic studies?

A. Heritability of Brain Structures in Healthy Persons

There are 24 studies reporting on heritability of brain structures in healthy persons. These studies are listed in Tables I and II, where the measured brain structures, the heritability rates of these brain structures or regions (denoted as h^2 or as ICC or as Risch's λ), and the method of calculating heritability (in superscript) is given, besides the sample size/mean age of the persons (twins, singletons, family members)/sex ratio/other variables, whenever available and mentioned in the articles are reported in the table as well, with the study ID and

TABLE I
STRUCTURAL BRAIN PHENOTYPES IN HEALTHY PERSONS WITH HIGH HERITABILITY
(h^2 BETWEEN 0.70 AND 0.99)

Brain structure	Heritability of brain structure (h^2, ICC or λ)	Sample size and age, gender, whenever different	Study ID
Cranial, intracranial volume	0.91	74 MZ, 71 DZ twin pairs (all M, 68–79 years)	Carmelli *et al.* (1998)[a]
	0.88	54 MZ, 58 DZ twin pairs, 34 siblings	Baaré *et al.* (2001a)[a]
	0.80	44 MZ, 40 DZ twin pairs (all M, 68–78 years)	Sullivan *et al.* (2001)[a]
	0.938	1881 individuals: 1330 genetically related (±60 years)	Atwood *et al.* (2004)[a]
Total brain volume	0.94	10 MZ (±34 years), 9 DZ (±23 years) twin pairs	Bartley *et al.* (1997)[a]
	0.97	23 MZ (12 M), 23 DZ (16 M) twin pairs, all reading disability, C (NRD) = 9 MZ (4 M), 9 DZ (4 M) twin pairs	Pennington *et al.* (2000)[b]
	0.90	54 MZ, 58 DZ twin pairs, 34 siblings	Baaré *et al.* (2001a)[a]
	0.99	12 MZ, 12 controls (all M)	White *et al.* (2002)[b]
	0.89	90 MZ (52 M, 38 F), 37 DZ (22 M, 15 F) twin pairs, 158 unrelated singletons (±11.5 years)	Wallace *et al.* (2006)[a]
	0.91	107 MZ and DZ twin pairs (±9 years)	Peper *et al.* (2009)[a]
Gray matter volume	0.82	54 MZ, 58 DZ twin pairs, 34 siblings	Baaré *et al.* (2001a)[a]
	0.99	12 MZ, 12 controls (all M)	White *et al.* (2002)[b]
	0.82	54 MZ, 58 DZ twin pairs, 34 siblings	Posthuma *et al.* (2002b)[b]
	0.82	90 MZ (52 M, 38 F), 37 DZ (22 M, 15 F) twin pairs, 158 unrelated singletons (±11.5 years)	Wallace *et al.* (2006)[a]
	0.91	107 MZ and DZ twin pairs (±9 years)	Peper *et al.* (2009)[a]
	0.55–0.85	49 MZ, 65 DZ twin pairs (±28 years)	Hulshoff Pol *et al.* (2006)[a]

White matter volume/hyper-intensity	0.71	74 MZ, 71 DZ twin pairs (all M, 68–79 years)	Carmelli *et al.* (1998)[a]
	0.88	54 MZ, 58 DZ twin pairs, 34 siblings	Baaré *et al.* (2001a)[a]
	0.98	12 MZ, 12 controls (all M)	White *et al.* (2002)[b]
	0.97	54 MZ, 58 DZ twin pairs, 34 siblings	Posthuma *et al.* (2002b)[b]
	0.85	90 MZ (52 M, 38 F), 37 DZ (22 M, 15 F) twin pairs, 158 unrelated singletons (± 11.5 years)	Wallace *et al.* (2006)[a]
	0.91	107 MZ and DZ twin pairs (±9 years)	Peper *et al.* (2009)[a]
	0.69–0.82	49 MZ, 65 DZ twin pairs (± 28 years)	Hulshoff Pol *et al.* (2006)[a]
Calosal vol.			
Size, shape	0.73, 0.90	5 MZ twin pairs, 10 unrelated controls	Oppenheim *et al.* (1989)[b]
Length, area, AC-PC distance	0.89, 0.906, 0.705	7 MZ twin pairs	Biondi *et al.* (1998)[a]
Area	$p = 0.02$	10 MZ twin pairs	Tramo *et al.* (1998)[b]
CC	0.79	44 MZ, 40 DZ twin pairs (all M, ± 68–78 years)	Sullivan *et al.* (2001)[a]
CC	0.94	14 MZ, 12 DZ twin pairs (±27 years)	Scamvougeras *et al.* (2003)[b]
CC	0.90	34 MZ, 37 DZ (all M, ± 76 years)	Pfefferbaum *et al.* (2004)[a]
CC	0.85	90 MZ, 37 DZ twin pairs, 158 unrelated singletons (± 11.5 years)	Wallace *et al.* (2006)[a]
CC	0.93	107 MZ and DZ twin pairs (±9 years)	Peper *et al.* (2009)[a]
Midline structures	0.83	107 MZ and DZ twin pairs (±9 years)	Peper *et al.* (2009)[a]
Cau, put, thal	0.84, 0.75, 0.75	12 MZ, 12 controls (all M)	White *et al.* (2002)[b]
Cau	0.80	90 MZ, 37 DZ twin pairs, 158 unrelated singletons (± 11.5 years)	Wallace *et al.* (2006)[a]
Middle frontal volume	0.90–0.95	10 MZ, 10 DZ twin pairs (50–70 years)	Thompson *et al.* (2001)[b]
FL, TL, PL	0.84, 0.88, 0.86	90 MZ, 37 DZ twin pairs, 158 unrelated singletons (± 11.5 years)	Wallace *et al.* (2006)[a]

(*continued*)

TABLE I (*continued*)

Brain structure	Heritability of brain structure (h^2, ICC or λ)	Sample size and age, gender, whenever different	Study ID
Cerebellum	0.88	43 MZ, 37 DZ twin pairs, 34 siblings	Posthuma *et al.* (2000)[a]
	0.99	12 MZ, 12 controls (all M)	White *et al.* (2002)[b]
Subcortex	0.70	23 MZ (12 M), 23 DZ (16 M) twin pairs, all reading disability, C (NRD) = 9 MZ (4 M), 9 DZ (4 M) twin pairs	Pennington *et al.* (2000)[b]
Lateral ventricle	0.83	34 MZ, 37 DZ (all M, ± 76 years)	Pfefferbaum *et al.* (2004)[a]
Cerebral ventricle	0.83	18 MZ, 21 DZ twin pairs (±37 years)	Reveley *et al.* (1984)[a,b,c]

Summary of published studies by August 2009 (PubMed and Medline citations) on the heritability of brain structural phenotypes including study results from healthy monozygotic (MZ) and dizygotic (DZ) twins, singletons (sibs), and in some studies healthy controls (C). Heritability is denoted as h^2 or as intraclass correlation factor (ICC) or as Risch's λ. A higher ICC for MZ twins then for DZ twins indicates a genetic component to these structures, but large confidence intervals indicate that the similarities measured are a rough estimates of the degree to which a brain structure is genetically controlled. Another way of denoting the heritability is by λ_s = Risch's λ = a measure of heritability. For Risch's λ, $\lambda_s = 1$ is the expected value for nonfamilial traits, whereas $\lambda_s > 2.0$ is considered evidence for familiality. h^2 or ICC or Risch's λ = heritability rate.

TB, total brain volume; WM, total white matter volume; GM, total gray matter volume; WM, white matter volume; GM, gray matter volume; CC, corpus callosum; AC-PC distance, the distance between anterior and posterior commissure; CB, cerebellar volume; FL, frontal lobe; TL, temporal lobe; PL, parietal lobe; OL, occipital lobe; OC, occipital cortex; TH, temporal horn volume; LV, lateral ventricle volume; TV, third ventricle; HC, hippocampal volume; Parahip, parahippocampal gyrus; Cau, caudate nucleus; Thal, thalamus; SS or SL, sulcus shape or sulcus length; put, putamen; Hpt, hypothalamus; PHC gyrus, parahippocampal gyrus; MZ, monozygote; DZ, dizygote; M, male, F, female; Sibs, siblings; n.a., not available; n.s., not significant.

[a]SEM, structural equitation modeling., [b]Falconer method., [c]Analysis of variance estimate of heritability.

TABLE II

Structural Brain Phenotypes in Healthy Persons with Moderate Heritability (h^2 between 0.40 and 0.70), Low Heritability (h^2 Between 0.00 and 0.40) and No Significant Heritability ($h^2 = 0$ and Denoted as n.s.)

Brain structure	Heritability of brain structure (h^2, ICC or λ)	Sample size	Study ID
Frontal lobe (L, R), Temporal lobe (L, R), Parietal lobe (L, R) Occipital lobe (L, R)	0.66, 0.56 0.59, 0.41 (n.s.) 0.41 (n.s.), 0.47 0.28 (n.s.), 0.69	72 MZ, 67 DZ twin pairs (±71 years)	Geschwind *et al.* (2002)[a,b,c]
Total brain volume	0.66	9 MZ, 10 DZ twin pairs (± 23 years)	Wright *et al.* (2002)[b]
	0.61	72 MZ, 67 DZ twin pairs (±71 years)	Geschwind *et al.* (2002)[a,b,c]
Hippocampus	0.40	44 MZ, 40 DZ twin pairs (all M, 68–78 years)	Sullivan *et al.* (2001)[b]
Neocortex	0.56	23 MZ (12 M), 23 DZ (16 M) twin pairs, all reading disability, C (NRD) = 9 MZ (4 M), 9 DZ (4 M) twin pairs	Pennington *et al.* (2000)[a]
Cerebral Cortex	0.30–0.57	107 MZ, 47 DZ twin pairs, siblings of twins 64, singletons 228	Schmitt *et al.* (2008)[b]
WM hyperintensity	0.553	1881 individuals: 1330 genetically related (± 60 years)	Atwood *et al.* (2004)[b]
Planum Temporale asymmetry	0.612	27 MZ, 12 DZ win pairs (± 27 years)	1) Eckert *et al.* (2002)[a]
	MZ cc:0.21, MZ disc: 0.41	10 MZ concordant and 10 MZ discordant twin pairs for handedness (±26 and 22 years)	2) Steinmetz *et al.* (1995)[a]
Brain Parenchyma	0.62	74 MZ, 71 DZ twin pairs (all M, ± 68–79 years)	Carmelli *et al.* (1998)[b]
Temporal Horn	0.60	44 MZ, 40 DZ twin pairs (all M, 68–78 years)	Sullivan *et al.* (2001)[b]
CC perimeter	0.620	7 MZ twin pairs	Biondi *et al.* (1998)[d]
Cerebellum	0.49	90 MZ, 37 DZ twin pairs, 158 unrelated singletons (± 11.5 years)	Wallace *et al.* (2006)[b]

(*continued*)

TABLE II (*continued*)

Brain structure	Heritability of brain structure (h^2, ICC or λ)	Sample size	Study ID
Other (n.s. or low h^2):			
SS or SL	0.377 (n.s.)	27 MZ, 12 DZ twin pairs (±6–16 years)	Eckert *et al.* (2002)[a]
LV	0.31	90 MZ, 37 DZ twin pairs, 158 unrelated singletons (± 11.5 years)	Wallace *et al.* (2006)[b]

Summary of published studies by August 2009 (PubMed and Medline citations) on the heritability of brain structural phenotypes including study results from healthy monozygotic (MZ) and dizygotic (DZ) twins, singletons (sibs), and in some studies healthy controls (C). Heritability is denoted as h^2 or as intraclass correlation factor (ICC) or Risch's λ. A higher ICC for MZ twins then for DZ twins indicates a genetic component to these structures, but large confidence intervals indicate that the similarities measured are a rough estimates of the degree to which a brain structure is genetically controlled. Another way of denoting the heritability is by λ_s = Risch's λ = a measure of heritability. For Risch's λ, $\lambda_s = 1$ is the expected value for nonfamilial traits, whereas $\lambda_s > 2.0$ is considered evidence for familiality. h^2 or ICC or Risch's λ = heritability rate.

TB, total brain volume; WM, total white matter volume; GM, total gray matter volume; WM, white matter volume; GM, gray matter volume; CC, corpus callosum; AC-PC distance, the distance between anterior and posterior commissure; CB, cerebellar volume; FL, frontal lobe; TL, temporal lobe; PL, parietal lobe; OL, occipital lobe; OC, occipital cortex; TH, temporal horn volume; LV, lateral ventricle volume; TV, third ventricle; HC, hippocampal volume; Parahip, parahippocampal gyrus; Cau, caudate nucleus; Thal, thalamus; SS or SL, sulcus shape or sulcus length; put, putamen; Hpt, hypothalamus; PHC gyrus, parahippocampal gyrus; MZ, monozygote; DZ, dizygote; M, male, F, female; Sibs, siblings; n.a., not available; n.s., not significant.

[a]Falconer method., [b]SEM, structural equitation modeling., [c]Analysis of variance estimate of heritability., [d]Visual comparison of the 3D surface rendering models of the brains and manual measurements of the AC-PC distance by six blinded observers.

year of publication. Magnetic resonance imaging (MRI scans) brain scans were obtained from MZ and dizigotic (DZ) healthy twin pairs. The inclusion of monozygotic, dizygotic, singletons, and family members, but also unrelated controls is done to assess the extent of genetic control over these structures. Within pair similarities for certain brain structures can be compared for each group, for example within pair similarity for brain structures with high heritability rates such as the total brain volume increases as pair members are more closely related genetically (monozygotic > siblings (included dizygotic) > unrelated controls). In one study healthy controls were also included (Oppenheim *et al.*, 1989). In five studies, unrelated and related singletons were included as well (Baaré *et al.*, 2001a; Posthuma *et al.*, 2000, 2002b; Schmitt *et al.*, 2008; Wallace *et al.*, 2006), and in one study (Atwood *et al.*, 2004) family members were also included. In three studies the heritability rates were investigated only in a male population (Carmelli *et al.*, 1998; Eckert *et al.*, 2002; Sullivan *et al.*, 2001), and the remaining studies consisted of a mixed population. The age range for most of the studies was between 23 and 80 years, except for one study (Peper *et al.*, 2009) in which 9-year-old children were included.

There were 37 structures measured in total. Some structures were measured in multiple studies, such as the total brain volume (six studies), total cranial/intracranial volume (four studies), white matter (seven studies), white matter hyperintensity (two studies), gray matter (six studies), corpus callosum (eight studies), cerebellar volume (three studies), frontal/temporal/parietal lobe (two studies), lateral ventricle (two studies), amygdalar and caudate (two studies), sulcus shape/length (two studies), planum temporal asymmetry (two studies), and ventricular volume (three studies). The rest of the structures were measured in a single study. In one study (Hulshoff Pol *et al.*, 2006) gray matter and white matter volume was measured for different brain parts such as superior frontal, medial frontal, postcentral gyrus, posterior cingulated, and Heschl's gyrus for the left and the right part. In Table I, the heritability rates ranging from 0.55 to 0.85 for gray matter and 0.69–0.82 for white matter volume are given. In another study (Schmitt *et al.*, 2008) various parts of the cerebral cortex were investigated and in Table II heritability rates between 0.30 and 0.57 for the cerebral cortex are recorded.

1. *Results on the Heritability of the Reviewed Studies*

Gross brain structures show higher heritability rates than specific structures. High heritability rates ranging from 0.70 to 0.99 are characteristic to intracranial volume, whole brain volume, total gray matter volume, total white matter volume, calosal volume, middle frontal volume, midline structures, and the volumes of both the right and left hemispheres (Table I). Studies including DZ, resulting in a full twin study (Bartley *et al.*, 1997; Carmelli *et al.*, 1998; Pennington *et al.*, 2000), reported heritability of >0.90 for total brain volume. A moderate heritability rate was found on the four lobes investigated, such as frontal, temporal, occipital, and parietal lobe, ranging from 0.41 to 0.70 (Geschwind *et al.*, 2002), but conflicting results were reported on the lobes by Wallace *et al.* (2006) for the frontal, temporal and parietal lobe, resulting in high h^2 scores between 0.84 and 0.88. Moderate heritability rates are also reported for temporal horn volume, brain parenchyma, hippocampus, neocortex, white matter hyperintensity, and planum temporal asymmetry ranging from 0.60 to 0.63. Studies showing a moderate or a low heritability of brain structures are listed in Table II.

2. *Effects of Gender, Age, Handedness, and Development in Time*

Some studies have investigated variables that could have an influence on heritability rates. In a study by Steinmetz *et al.* (1995), the effect of gender on measuring the midsagittal area of the corpus callosum and total brain volume in 120 healthy young adults was shown, finding that the forebrain volume-adjusted size of the corpus callosum was larger in women than in men and that handedness had no effect on this measurement. However, an effect of handedness on cerebral asymmetry was found in two studies (Eckert *et al.*, 2002; Steinmetz *et al.*, 1995) reporting a lack of cerebral asymmetry in persons who were left handed. This might be explained by the hypothesis that early "epigenetic" events that take place during embryogenesis may contribute to a significant variability in the development of the anatomofunctional laterality of the cerebral hemispheres. In a study by Peper *et al.* (2009), children ($n = 47$) with signs of secondary sexual characteristics (based on Tanner stages) were compared to children without these signs and the heritability of brain structures was estimated using MRI scans. They found high heritability for most gross brain structures (see Table I) and regional brain structures. They also found that the onset of secondary sexual characteristics of puberty were associated with decreased frontal and parietal gray matter densities that accompany the transition of our brains from childhood into adulthood, still showing high heritability for brain structures such as global brain volumes, white matter density in fronto-occipital and superior longitudinal fascicles and gray matter density of (pre)frontal and temporal areas. Only one study (Pfefferbaum *et al.*, 2004) investigated the heritability rates examined over time and found no evidence for new genetic variance on the h^2 scores of corpus callosum and lateral ventricles recalculated after 4 years. The heritability of surface morphology, which is formed later in development and assumed to be under more environmental influence than the deeper structures (formed earlier in life and under more genetic control) indicate that sulcal depth effect was stronger for related twins. These results show that the shape of the deeper sulci is more strongly predetermined than that of superficial sulci, for example, sulcal depth having an influence on sulcal similarity. Also, subjects with similar brain shape have more similar patterns of their sulci, suggesting both features are the result of ontogenetically related processes (Bartley *et al.*, 1997; Biondi *et al.*, 1998;

Lohmann *et al.*, 1999; White *et al.*, 2002). This anatomical similarity was also found for the corpus callosum shape and size (Oppenheim *et al.*, 1989).

3. *Conflicting Results on Some Brain Structures*

Two studies (Posthuma *et al.*, 2000; White *et al.*, 2002) report high h^2 rates for cerebellar volume and one study (Wallace *et al.*, 2006) reports a moderate h^2 rate. This might be due to the age difference of the participants, and thus, the ongoing development of the brain. In the first two studies, adults were included, whereas the latter study, included adolescents (mean age of 11.5 years). A low heritability rate for lateral ventricle volume was found in one study (Wallace *et al.*, 2006), but a high h^2 score was reported in another (Pfefferbaum *et al.*, 2004), and an h^2 of 0.00 is recorded by Wright *et al.* (2002). The differences in these studies might be due to the differences in age of the participants, but could also be due to the differences in sample size. The mean age of the participants in the study by Wallace *et al.* (2006) was 11.5 years, reflecting results of brains in development. Additionally, the sample size in the study by Wright *et al.* (2002) is small compared to the bigger sample sizes of the other two studies. Conflicting results also exist on the heritability of sulcus shape and length, with one study indicating moderate heritability rate (Bartley *et al.*, 1997) and not reaching significance in another (Eckert *et al.*, 2002). There is a difference in the mean age between the studies, resulting in an older mean age (23 years) in the study by Bartley *et al.* (1997) and younger (6–16 years) in the study by Eckert *et al.* (2002). However, it is unclear what might be of influence in the differences between these studies, because different statistical heritability estimates were used in both. Other conflicting results are found for the ventricle volume, having a high h^2 rate in one study (Reveley *et al.*, 1984) and not being significant in other studies (Baaré *et al.*, 2001a; White *et al.*, 2002). In the study by Reveley *et al.* (1984), three statistical ways of calculating heritability rates were used, whereas the other two studies used one of the methods (both different from each other).

The differences between the studies with conflicting results might be a result of the differences in sample size, age, gender of the population (only male participants vs. mixed group), statistical method applied to calculate the heritability rates and the way measurements of the brain structures is done. As for the latter, some studies used the manual measurement of brain structures (region or VOI) and others automated (VBM, deformation based morphometry).

Seven studies used the Falconer method of heritability calculation and 15 studies used the structural equitation modeling (SEM), in which the unique environmental factors are also included. Two studies (Geschwind *et al.*, 2002; Reveley *et al.*, 1984) used all three methods of heritability calculation, reaching significance in both studies for the three methods.

Heritability estimates show considerable differences between gross and specific brain structures. High heritability is characteristic for gross brain structures, such as intracranial volume, whole brain volume, cerebellar volume, total gray matter and white matter volume, and calosal volume, with effect sizes in the high range. Moderate effect sizes have been found for structures such as the heritability of the different lobes, sulcus shape or length, white matter hyperintensity, hippocampal volume, and other brain structures like brain parenchyma, middle frontal regions, temporal horn and planum temporal asymmetry. The highly heritable gross brain structures are also the brain structures that are examined in multiple studies. The volume measurements of these brain structures and thus, the heritability rates, are easier to calculate than the specific, middle brain, deeper or small brain structures in both the manual and automated quantitative measurement techniques. From the reviewed literature, these brain structures with high and moderate heritability rates seem to be useful as endophenotypes.

B. Heritability of Brain Structures in Patients with Schizophrenia Compared to Healthy Controls

Heritability calculations of the brain structures of patients with schizophrenia compared to healthy controls is very complicated, due to the impact of disease related factors on brain changes, together with the effect of medication intake and the use of drugs (Kanaan *et al.*, 2009; Rais *et al.*, 2008; van Haren *et al.*, 2007). Some studies show an increase in brain structure volumes and others a decrease. However, use of medication does not account for the total changes in the brain structures of patients with schizophrenia, since there are also studies showing brain changes in medication naïve patients (Cahn *et al.*, 2001; Okugawa *et al.*, 2007; Weinberger, 1995). Using twins concordant and discordant for schizophrenia in these studies would help elaborate the disease's effect on the brain changes to a certain degree.

Until now, there have been nine studies published on the heritability of brain structures in patients with schizophrenia compared to healthy controls. These studies are listed in Table III. The studies report on the heritability rates of brain structures in twins concordant and discordant for schizophrenia, monozygotic, and DZ, compared to healthy twins. These studies included adult patients, females, and males. Some brain structures have been studied in multiple studies, such as the total brain volume (four studies), gray matter volume (two studies), hippocampal volume (four studies), and the rest of the brain structures listed on the table in single studies.

High heritability for the intracranial volume, total gray and white matter volume, ranging from 0.72 to 0.89 was found (Table III). The overall gray matter volume was also found to be highly heritable in monozygote twins, both in the control and in the group of patients with schizophrenia. The heritability of total brain volume was found to be high in two studies (Hulshoff Pol *et al.*, 2004; Rijsdijk *et al.*, 2005) and high in patients with schizophrenia, but moderate in healthy controls in another study (Narr *et al.*, 2002). In one study (Rijsdijk *et al.*, 2005), a low heritability rate was found for the hippocampus, third and lateral ventricle.

TABLE III

HERITABILITY OF STRUCTURAL BRAIN PHENOTYPES IN PERSONS WITH SCHIZOPHRENIA COMPARED TO HEALTHY CONTROLS

Brain structure	Heritability of brain structure (h^2, ICC or λ)	Sample size	Study ID
Intracranial or cranial volume	(1) SCZ: MZ:0.93, DZ:0.45 C: MZ:0.86, DZ: 0.35	(1) 15 MZ, 15 same sex DZ disc for SCZ, 29 healthy twin pairs	(1) Baare *et al.* (2001b)[a]
	(2) 0.89	(2) 11 MZ, 11 DZ disc for SCZ, 11 MZ and 11 DZ C twin pairs	(2) Hulshoff Pol *et al.* (2004)[a]
TB volume	(1) SCZ: MZ: 0.86, DZ: 0.06 C: MZ: 0.93, DZ: 0.36	(1) 15 MZ, 15 same sex DZ disc for SCZ, 29 healthy twin pairs	(1) Baare *et al.* (2001b)[a]
	(2) 0.86	(2) 11 MZ, 11 DZ disc for SCZ, 11 MZ and 11 DZ C twin pairs	2) Hulshoff Pol *et al.* (2004)[a]
	(3) 0.88	(3) 14 MZ cc for SCZ and 10 MZ disc for SCZ, 17 MZ C pairs, 22 disc sibling pairs, 3 cc sibling pairs, 114 healthy C	(3) Rijsdijk *et al.* (2005)[a]
	(4) 0.66	(4) 9 MZ, 9 DZ disc for SCZ, 14 MZ and 13 DZ healthy twin pairs	(4) Brans *et al.* (2008)[a]
GM volume	(1) SCZ: MZ: 0.73, DZ:0.05 C: MZ: 0.83, DZ = 0.32	(1) 20 MZ, 20 DZ twin pairs with SCZ, 20 MZ and 20 DZ healthy twin pairs (± 48 years)	(1) Cannon *et al.* (2002)[b]
	(2) 0.72	(2) 11 MZ, 11 DZ disc for SCZ, 11 MZ and 11 DZ C twin pairs	(2) Hulshoff Pol *et al.* (2004)[a]
WM volume	0.72	11 MZ, 11 DZ disc for SCZ, 11 MZ and 11 DZ C twin pairs	Hulshoff Pol *et al.* (2004)[a]
Frontal lobe	(1) FL; SCZ: MZ:0.84, DZ: 0.26 C: MZ: 0.93, DZ: 0.00	(1) 15 MZ, 15 same sex DZ disc for SCZ, 29 healthy twin pairs	(1) Baare *et al.* (2001b)[a]
Parietal lobe	(2) FL: 0.76, TL: 0.79	(2) 9 MZ, 9 DZ disc for SCZ, 14 MZ and 13 DZ healthy twin pairs	(2) Brans *et al.* (2008)[a]

(continued)

TABLE III (*continued*)

Brain structure	Heritability of brain structure (h^2, ICC or λ)	Sample size	Study ID
Hippocampus	(1) SCZ: MZ: 0.68, DZ: 0.00 C: MZ: 0.73, DZ: 0.00	(1) 15 MZ, 15 same sex DZ disc for SCZ, 29 healthy twin pairs	(1) Baare *et al.* (2001b)[a]
	(2) SCZ: 0.42, C: 0.71	(2) 7 MZ cc and 16 MZ disc and 28 DZ disc twin pairs for SCZ, 28 MZ and 26 DZ healthy control twin pairs	(2) van Erp *et al.* (2004)[a]
	(3) 0.36	(3) 14 MZ cc and 10 MZ disc for SCZ	(3) Rijsdijk *et al.* (2005)[a]
	(4) L:3.10, R: 2.33 (λ_s)	(4) SCZ: 169 patients, unaffected siblings: 183, controls: 221	(4) Goldman *et al.* (2007)
Corpus callosum	(1) SCZ: MZ:0.91, DZ:0.33 C: MZ: 0.54, DZ: 0.74	(1) 10 MZ, 10 DZ twin pairs disc for SCZ and 10 MZ, 10 DZ unaffacted co twin pairs	(1) Narr *et al.* (2002)[b]
	(2) L: 4.93, R: 8.50 (λ_s)	(2) SCZ: 169 patients, unaffected siblings: 183, controls: 221	(2) Goldman *et al.* (2007)
Third ventricle	(1) SCZ: MZ:0.36, DZ: 0.17 C: MZ: 0.71, DZ: 0.27	(1) 15 MZ, 15 same sex DZ disc for SCZ, 29 healthy twin pairs	(1) Baare *et al.* (2001b)[a]
	(2) 0.33	(2) 14 MZ cc for SCZ and 10 MZ disc for SCZ, 17 MZ C pairs, 22 disc sibling pairs, 3 cc sibling pairs, 114 healthy C	(2) Rijsdijk *et al.* (2005)[a]
Lateral ventricle	(1) SCZ: MZ: 0.67, DZ:0.18 C: MZ: 0.73, DZ: 0.00	(1) 15 MZ, 15 same sex DZ disc for SCZ, 29 healthy twin pairs	(1) Baare *et al.* (2001b)[a]
	(2) 0.05	(2) 14 MZ cc for SCZ and 10 MZ disc for SCZ, 17 MZ C pairs, 22 disc sibling pairs, 3 cc sibling pairs, 114 healthy C	(2) Rijsdijk *et al.* (2005)[a]
	(3) L: 0, R: 0 (λ_s)	(3) SCZ: 169 patients, unaffected siblings: 183, controls: 221	(3) Goldman *et al.* (2007)

Hypothalamus	MZ: ICC = 0.58, $p < 0.05$, DZ: 0.07. Correction for TB→n.s.	11 MZ, 11 DZ same sex twin pairs, disc for SCZ and 11 MZ, 11 DZ same sex healthy twin pairs	Koolschijn *et al.* (2007)[a]
PHC gyrus	SCZ: MZ: 0.34m DZ: 0.37 C: MZ: 0.70, DZ: 0.10	15 MZ, 15 same sex DZ disc for SCZ, 29 healthy twin pairs	Baare *et al.* (2001b)[a]
Dorsal striatum	L: 0, R: 0.87 (λ_s)	SCZ: 169 patients, unaffected siblings: 183, controls: 221	Goldman *et al.* (2007)

Summary of published studies by August 2009 (PubMed and Medline citations) on the heritability of brain structural phenotypes with study results from twins concordant (cc) and discordant (disc) for schizophrenia (SCZ), singletons, and healthy controls (C). Heritability is denoted as h^2 or as intraclass correlation factor (ICC). A higher ICC for monozygotic (MZ) twins then for dizygotic (DZ) twins indicates a genetic component to these structures, but large confidence intervals indicate that the similarities measured are a rough estimates of the degree to which a brain structure is genetically controlled.

Another way of denoting the heritability is by λ_s = Risch's λ = a measure of heritability. For Risch's λ, $\lambda_s = 1$ is the expected value for nonfamilial traits, whereas $\lambda_s > 2.0$ is considered evidence for familiality. h^2 or ICC or Risch's λ = heritability rate.

TB, total brain volume; WM, total white matter volume; GM, total gray matter volume; WM, white matter volume; GM, gray matter volume; CC, corpus callosum; AC-PC distance, the distance between anterior and posterior commissure; CB, cerebellar volume; FL, frontal lobe; TL, temporal lobe; PL, parietal lobe; OL, occipital lobe; OC, occipital cortex; TH, temporal horn volume; LV, lateral ventricle volume; TV, third ventricle; HC, hippocampal volume; Parahip, parahippocampal gyrus; Cau, caudate nucleus; Thal, thalamus; SS or SL, sulcus shape or sulcus length; put, putamen; Hpt, hypothalamus; PHC gyrus, parahippocampal gyrus; MZ, monozygote; DZ, dizygote; M, male; F, female; Sibs, siblings; n.a., not available; n.s., not significant.

[a]SEM, structural equitation modeling.

[b]Falconer method.

Low heritability rates were also found in the group of DZ twins discordant for schizophrenia (Narr *et al.*, 2002), for the overall gray matter volume in control DZ twins and in DZ twins with schizophrenia. There was no significant heritability found for the parahippocampal gyrus, third and lateral ventricle (Baaré *et al.*, 2001b) or for the hypothalamus (Koolschijn *et al.*, 2007). The study by Brans *et al.* (2008) reports on whether genetic (heritability) and/or environmental factors are associated with progressive brain volume changes in schizophrenia. They found that significant decreases over time in whole brain, frontal, and temporal lobe volume in patients with schizophrenia and in their unaffected co-twins compared with control twins revealed significant additive influences on the correlations between schizophrenia liability and progressive whole brain (0.66), temporal (0.79), and frontal (0.76) lobe changes.

The samples in these studies consist of both healthy MZ and DZ twins and twins concordant and discordant for schizophrenia, with the exception of three studies (Baaré *et al.*, 2001b; Hulshoff Pol *et al.*, 2004; Koolschijn *et al.*, 2007) where twins concordant for schizophrenia were not included, resulting in a sample group where environmental factors played a more important role in the development of the disorder and a genotypic estimation of the true population genetic variance was not possible.

Three studies (Cannon *et al.*, 2002; Narr *et al.*, 2002) used the Falconer method of heritability estimates and four studies used the SEM method (Brans *et al.*, 2008; Hulshoff Pol *et al.*, 2004; Rijsdijk *et al.*, 2005; Van Erp *et al.*, 2004). The study by Koolschijn *et al.* (2007) used the intraclass correlation factor (ICC) to estimate the heritability rates of the hypothalamus. One study, by Goldman *et al.* (2007), used ICC and also Risch's relative risk for siblings (λ_s) for heritability measurements of each region. Risch's relative risk for siblings (λ_s), was calculated to quantitatively measure the likelihood that siblings of patients with volumetric reductions or increases would express the phenotype themselves. Because this parameter depends on the presence or absence of the phenotype, cutoffs were defined on the basis of the distribution of the volumes in patients.

One complicating factor in the heritability estimates of brain structures in twin studies is that the twin discordant for schizophrenia also shows, to a milder degree, brain changes seen in their co-twins with schizophrenia. The unaffected twins show lower brain volume, lower hemispheric volumes and/or frontal lobe gray matter volume decrease. These observations may suggest the involvement of genetic factors in cortical gray matter deficits in patients with schizophrenia (Baaré *et al.*, 2001b; Reveley *et al.*, 1984; Suddath *et al.*, 1990). As suggested by Cannon *et al.* (2002), it is important that these studies include an evaluation of the regional cortical variation and that individual differences in gyral/sulcal patterning and shape are taken into account. It is not clear to what extent these changes are similar to the changes in their co-twins who have schizophrenia. A distinction between these possibly genetic differences and the disease related changes in the brain should be done by isolating the topography of nongenetic, disease related differences in cortical gray matter volume, as suggested by Cannon *et al.* (2002).

C. Heritability of Brain Structures in Nonhuman Primates (Monkeys)

Studies in pedigrees, where there is an extensive control of the environmental factors, are particularly useful in heritability calculations. Studies in NHPs are therefore a good model for genetic control and could be useful in heritability calculations as mentioned previously.

There are five studies published on the heritability of brain structures in NHPs (e.g., monkeys). These studies are listed in Table IV. Results of these studies strongly support findings from human studies showing the heritability of gross brain structures in the high effect size range, such as whole brain volume, cerebral volume, and cranial size. The moderate effect size range for heritability was reported for cranial capacity, sulcus length, brain weight, hippocampus and surface area, and gray matter. The number of studies reporting on heritability of brain structures in NHPs is limited and the sample sizes of monkeys included in these studies is small. There are only two studies (Mahaney *et al.*, 1993; Rogers *et al.*, 2007) in which the same monkey type was included, but these studies did not look at the same brain structure, which makes comparison difficult. Only the study by Fears *et al.* (2009) used a big sample size of vervet monkeys and calculated heritability estimates using manual tracings as well as automated tracings of brain structures. In this study, high heritability rates are found for gross brain structures, such as total brain volume, cerebral volume and corpus callosum. In contrast to other NHP heritability estimations of the hippocampus, this study showed a high heritability rate for the hippocampus. The findings from the NHP studies support the findings in human twin studies.

Table IV shows that samples consist of different types of monkeys, making it difficult to compare the conflicting results between studies on the same brain structure, such as the hippocampus, where Lyons *et al.* (2001) find moderate heritability rates and Fears *et al.* (2009) find a high heritability rate. There is also a big difference in the sample sizes.

One other complicating factor is the method of measuring the volume of brain structures (manual vs. automated or both), particularly, the calculation of heritability of brain structures across the different studies. The method used by Lyons *et al.* (2001) consists of obtaining MRI scans of Guyanese monkeys and quantitative estimates of the heritability rates were generated by using the one-way ANOVA as used to evaluate paternal half siblings effect. Rogers *et al.* (2007) used the VBM approach to look at the regional differences in the genetic effects on brain structure and calculated the heritability estimates. The study by Cheverud *et al.* (1990) shows that brain size was defined as endocranial capacity

TABLE IV

HERITABILITY OF STRUCTURAL BRAIN PHENOTYPES IN NONHUMAN PRIMATES

Brain structure	Heritability of brain structure (h^2, ICC or λ)	Sample size and monkey type	Study ID
Brain			
(1) Weight	(1) 0.409, $p < 0.005$	(1) Baboons: 601; 307 assigned to 25 pedigrees (2–49 members), 294 individual animals	(1) Mahaney *et al.* (1993)
(2) Size Cranial capacity	(2) 0.75, $p < 0.005$ 0.60, $p = 0.02$	(2) Rhesus Macaques; 179 skeletons, measuring endocranial capacity and using mother offspring regression and SDS method	(2) Cheverud *et al.* (1990)
Total brain volume	(1) 0.82 ($p = 0.00022$) (2) 0.99 (SE = 0.06)	(1) Baboons (papio hamadryas): 109 MRI Scans (VBM approach) (2) Vervet research colony: 357 (>2 years of age)	(1) Rogers *et al.* (2007) (2) Fears *et al.* (2009)
Gray matter volume	0.67, $p = 0.01$	Baboons (papio hamadryas): 109 MRI Scans (VBM approach)	Rogers *et al.* (2007)
Cerebral volume	(1) 0.86, $p = 0.0006$ (2) 0.98 (SE = 0.09)	(1) Baboons (papio hamadryas): 109 MRI Scans (VBM approach) (2) Vervet research colony: 357 (>2 years of age)	(1) Rogers *et al.* (2007) (2) Fears *et al.* (2009)
Surface area of the cerebrum	0.673, $p = 0.0069$	Baboons (papio hamadryas): 109 MRI Scans (VBM approach)	Rogers *et al.* (2007)
Corpus callosum	0.87 (SE = 0.07)	Vervet research colony: 357 (>2 years of age)	Fears *et al.* (2009)
Hippocampus	(1) 0.54 ($F = 2.58$, $p = 0.02$) (2) 0.95 (SE = 0.07)	(1) Guyanese (saimiri sciures): 39 MRI scans (1 way ANOVA) (2) Vervet research colony: 357 (>2 years of age)	(1) Lyons *et al.* (2001) (2) Fears *et al.* (2009)
Sulcus length	0.31, $p < 0.05$	Rhesus Macaques; 179 skeletons, measuring endocranial capacity and using mother offspring regression and SDS method	Cheverud *et al.* (1990)

Summary of published studies by August 2009 (PubMed and Medline citations) on the heritability results of brain structural phenotypes in nonhuman primates. Heritability is denoted as h^2.

SDS, symmetric differences squared method, SE, standard error, MRI, magnetic resonance imaging; VBM, voxel based morphometry method.

and was measured by filling the cranium with mustard seed, then pouring the enclosed seed into a graduated cylinder. The cube root of cranial capacity was used in statistical analysis to better fit the data with statistical models. The heritability was estimated by standard mother–offspring regression (Falconer and Mackay, 1996; Falconer *et al.*, 1981) and a slight modification of the symmetric-differences-squared (SDS) method (Bruckner and Slanger, 1986). In the study by Mahaney *et al.* (1993), the heritability estimates, specifically of the brain weight, were measured by quantitative genetic analyses on available loge-transformed organ weights. Fears *et al.* (2009) used both manual and deformation based volume measurements (automated) to measure volumes of brain structures.

Despite the difference in heritability rates of brain structures between the few studies published on this topic, the results are nonetheless comparable to the human studies. The newest publications on this subject in NHP (Fears *et al.*, 2009; Rogers *et al.*, 2007) have used big sample sizes and also comparable volume measurements on the MRI scans as seen in human studies. More research on heritability in NHPs will help us get more insight into the genetic control on brain structures and will create the potential for future genetic mapping of QTLs underlying the variance in the heritability of these brain structures.

III. Genes for Brain Structures in Healthy and Persons with Schizophrenia

The dramatically increased brain volume plays an important role in the origin of our species. In humans, brain volume is a quantitative trait with high heritability (Posthuma *et al.*, 2002b; Thompson *et al.*, 2001). Disturbances in brain development is one of the explanations for neuropsychiatric disorders such as schizophrenia (Weinberger, 1995), suggesting that developmental genes involved in the development of the brain may overlap with the causal genes, for example, potential candidate genes, for schizophrenia (Murray, 1994). However, no susceptibility genes with high effect sizes have been found for schizophrenia. Therefore, most studies have focused on the testing of specific genetic markers in a known candidate gene for association with the different endophenotypes. In this study, research using brain structures as neuroanatomical endophenotypes in healthy and patients with schizophrenia are reviewed. Studies on the association between specific genes and brain structures and between genes and brain morphological changes in schizophrenia will be reviewed.

A. Genes for Brain Structures in Healthy Persons

There are eight studies published on the genes for brain structures in healthy persons. These studies are listed in Table V. These studies looked at brain structures and specific genes (and alleles or SNPs, single nucleatide polymorphisms) for these brain structures and at the interaction between these genes.

TABLE V

GENES FOR STRUCTURAL BRAIN PHENOTYPES IN HEALTHY PERSONS

Genes (alleles)	Association with brain structures	Sample size	Study ID
BDNF (both Val66 Met allele)	(1) Hippocampal gray matter: Left HC GM: $p = 0.013$, $t = 2.24$, Right: $p < 0.001$, $t = 3.41$ (2) Hippocampal volume: ↓ 12,5%, df = 33.1, $F = 4.24$, $P = 0.048$	(1) 111 Persons (2) 36 persons	(1) Pezawas *et al.* (2004) (2) Bueller *et al.* (2006)
MCHP1 ASPM	(1) Brain volume: MCHP1: n.s. (df = 2, 112, $P = 0.20$), ASPM: n.s. (df = 2, 114, $p = 0.49$) (2) Cranial volume: MCHP1: ↑ volume in males with allele rs 1057090 ($R = 0.126$, $P = 0.016$	(1) 120 persons (2) 867 (387 M, 480 F)	(1) Woods *et al.* (2006) (2) Wang *et al.* (2008)
COMT	Gray matter density in anterior cingulate cortex: ↓ Gray Matter density	75 persons (persons at risk for SCZ)	McIntosh *et al.* (2007)
Neuroglin 1 (rs69994992)	White matter density: ↓ White Matter density ant limb of the internal capsula ($t = 4.41$, $p = 0.0028$ (T allele). ↓ Fractional anisotropy in C allele homozygotes	87 persons	McIntosh *et al.* (2008)
APOEε4	ε4 allele: ↓ Hippocampal volume	10 MZ twin pairs (±62.5 years)	Plassman *et al.* (1997)
RGS4 (A alleles)	Gray matter: RST gyrus: $Z = 3.49$, $p = 0.000$, Thalamus: $Z = 3.21$, $p = 0.001$, RVLPFC: $Z = 3.14$, $p = 0.001$, LIPL: $Z = 3.08$, $p = 0.001$ White matter: RVLPFC: $Z = 3.68$, $p = 0.000$, LVLPFC: $Z = 4.17$, $p = 0.000$, LHC: $Z = 3.77$, $p = 0.000$	106 persons	Buckholtz *et al.* (2007)

Summary of published studies by August 2009 (PubMed and Medline citations) on genes for brain structural phenotypes in healthy persons. The studies report on the brain structures that is investigated, the sample size, the results of the intereaction between genes (alleles) and brain morphological changes (significant or nonsignificant association (n.s.), a decrease (↓) or increase (↑) in brain volume/density in interaction with a gene or an allele) and on the (candidate) genes/alleles.

BDNF, brain derived neurotrophic factor; COMT, catechol-*o*-methyl transferase; NRG, neuroglin 1; MCPH1, microcephalin 1; ASPM, abnormal spindle-like microcephaly associated; RGS4, regulator of G protein signaling; C, controls; SCZ, schizophrenia; MZ, monozygotic; DZ, dizygotic; RVLPFC, right ventrolateral prefrontal cortex; LVLPFC, left ventrolateral prefrontal cortex; LIPL, left inferior parietal lobule; RST gyrus, right superior temporal gyrus.

In some of the studies the interaction of the specific genes with the variance in brain morphology is expressed in quantitative measures.

As presented in Table V, the genetics of heritability of brain structures indicate that in one study (Wang *et al.*, 2008) there is a significant association ($p = 0.016$) between the MCPH1 gene and intracranial volume, reaching bigger intracranial volume in patients with the rs1057090 allele and not reaching a significant association in another study (Woods *et al.*, 2006), not for the MCPH1 gene ($p = 0.20$) nor for the ASPM gene ($p = 0.49$). The heritability of gray matter density has been investigated in three studies. In the first study, there was a significant association for the right ($p < 0.001$) and left hippocampus ($p = 0.013$) for the *BDNF* gene for the Val66met allele carriers (Pezawas *et al.*, 2004). In the second study, there was a significant association in the right superior gyrus, temporal lobe, right prefrontal cortex, and left inferior parietal lobe ($p = 0.001$) (Buckholtz *et al.*, 2007) for the *RGS4* gene and specifically persons carrying the A allele. In the third study, there was a significant association in the anterior cingulate cortex for the *COMT* gene, specifically for persons carrying the *Val* gene (McIntosh *et al.*, 2007). The white matter density shows a significant association with two genes, namely, the *RGS4* gene (A allele carriers, $p = 0.000$) in the right and left ventral prefrontal cortex and left hippocampus in one study (Buckholtz *et al.*, 2007) and neuroglin 1 (rs699994992 allele carriers, $p = 0.00028$) with a white matter density decrease in anterior limb of the internal capsula in another study (McIntosh *et al.*, 2007). The hippocampal volume and the association with the APOEε4 gene shows HC volume decrease in one study (Plassman *et al.*, 1997) and of 11.2% reduction in volume in another study ($p = 0.048$) (Bueller *et al.*, 2006) for the *BDNF* gene Val66met allele carriers.

The samples consist of healthy individuals, not related, except for one study in which 10 MZ twin pairs were compared (Plassman *et al.*, 1997) and in another study, persons at risk for schizophrenia were included (McIntosh *et al.*, 2007). In the study by Woods *et al.* (2006), individuals from different ethnic backgrounds were included, but correcting the data for race, ethnicity or sex did not change the results. The sample in the study by Wang *et al.* (2008) consisted of 90% Chinese and 10% other ethnic minorities. The study by Buckholtz *et al.* (2007) consisted of Caucasian subjects of European descent. The sample in the study by McIntosh *et al.* (2007) consisted of persons at risk for schizophrenia.

The method used by the studies in Table III varies from one study to the other. In the study by Woods *et al.* (2006), heritability of brain volume related to genotype was estimated by using the coefficient of the genetic terms in the full fitted ANOVA model to compute the variance attributable to genotype. This variance was then divided by the total variance to estimate heritability. This method was also used by Bueller *et al.* (2006) and Pezawas *et al.* (2004). The study by Wang *et al.* (2008) used the linear regression model to do the single SNP association analysis. Buckholtz *et al.* (2007) used multimodal neuroimaging to investigate the impact on brain structure and function and looking at allelic variation. The study by McIntosh *et al.* (2007) imported the data into SAS (version 9.1, SAS Institute, Cary, North Carolina) and the effects of group status by COMT group interactions were investigated by analysis of variance.

B. Genes for Brain Structures in Patients with Schizophrenia

There are 23 studies where the genetics of variation on heritability of brain structures was investigated in patients with schizophrenia compared to healthy controls. These studies report on the association between specific genes and brain morphological changes in schizophrenia. Several candidate genes for schizophrenia have been studied and the association with specific brain morphological changes has been examined. There are multiple studies on the Brain Derived Neurotrophic Factor gene (*BDNF* gene) (five studies), Catecol-*O*-Methyl Transferase gene (*COMT* gene) (three studies), Disrupted in Schizophrenia 1 (DISC1) (two studies), APOE ε4 gene (two studies) and Interleukin-1 receptor antagonist gene (IL 1) (two studies). These studies are listed in Table VI. There are single studies on the Neuroglin 1 (*NRG1*), prion protein (*PP* gene), tumor necrosis factor receptor-II gene (*TNF-RII* gene), *NOTCH4* gene, *PCM1* gene, *Plexin B3* gene, *GAD1* gene, *PROHD* gene, regulation of G protein signaling (*RGS4*) and *neurotrophin* gene. These studies are listed in Table VII. The brain structures that are looked at in multiple studies include: the brain lobes (six studies), hippocampal volume (eight studies), ventricles (three studies), gray matter volume (seven studies), white matter volume (seven studies) and cerebrospinal fluid (CSF) volume (three studies). The following structures have been studied in a single study: cerebral asymmetry, intracranial volume, regional gray matter volume/density, prefrontal cortex, caudate, putamen, and cerebellum.

One of the most investigated genes is the *BDNF* gene. BDNF is a neurotrophic factor involved in the development and maintenance of the nervous system. It is neuroprotective and prevents cell loss in the cerebral cortex, the striatum, and the hippocampus (Angelucci *et al.*, 2005). The *BDNF* gene consists of several polymorphisms that affect the amino acid sequence and possibly the function of the protein. In patients with schizophrenia, lower concentrations of BDNF have been found in cortical areas and the hippocampus as well as in the blood serum. The *COMT* gene is thought to be involved in the degradation of dopamine. The BDNF allele, in particular the Met allele, was found to be associated with a decrease in hippocampal volume (Bueller *et al.*, 2006) and with frontal lobe and occipital lobe gray matter in two other studies (Ho *et al.*, 2006; Pezawas *et al.*, 2004). Szeszko *et al.* (2005) found an association in a post hoc analysis between hippocampal volume decrease for schizophrenia patients only and no association with the intracranial volume. Wassink *et al.* (1999) studied the BDNF alleles in 48 trios and found an association with the parietal lobe, but not with the temporal or

TABLE VI

MOST STUDIED GENES FOR STRUCTURAL BRAIN PHENOTYPES IN PATIENTS WITH SCHIZOPHRENIA COMPARED TO HEALTHY PERSONS

Genes (alleles)	Association of genes/alleles with brain structures	Sample size	Study ID
BDNF			
(1) BDNF	(1) Parietal lobe: $\chi = 7.90$, $p = 0.0049$, n.s. with FL, TL, OL, ventricles	(1) 48 trios	(1) Wassink *et al.* (1999)
(2) Val66Met vs val/val	(2) Hippocampus: $F = -3.22$, df = 42, $p = 0.002$ (SCZ + C); $F = 11.72$, df = 15, $p = 0.004$ (SCZ), Intracraniaal volume: $t = -1.24$, df = 42, $p = 0.221$ (SCZ + C)	(2) 19 SCZ, 25 C	(2) Szeszko *et al.* (2005)
(3) Met carriers	(3) GM volume in 2 cognitive domains: ↓ GM vol in TL: $F = 5.44$, $p = 0.02$ and OL: $F = 5.73$, $p = 0.02$	(3) 183 SCZ, 80 C	(3) Ho *et al.* (2006)
(4) Polymorphisms	(4) All brain structures in literature identified abnormal in SCZ: 111757G/C: larger frontal GM ($p < 0.001$) (SCZ), 270 C/T: totale caudate volume ($p = 0.006$) and GM in caudate ($p = 0.022$), 111757G/C + 270 C/T + -633T/A + Val66Met = n.s. for HC, n.s. for controls either with any brain structure except putamen ($p = 0.006$)	(4) 30 SCZ, 104 C	(4) Agartz *et al.* (2006)
(5) Val66Met	(5) ↓Frontal GM in Met allele carriers ($F = 4.20$, df = 1, 113, $p = 0.05$), ↑lateral ventricles and sulcal CSF ($t \geq 3.95$, df = 44, $p \leq 0.0003$), ↑ CSF in FL and TL ($F \geq 6.71$, df = 1, 113, $p \leq 0.01$), n.s. in PL and OL ($F \leq 1.84$, df = 1, 113, $p \geq 0.18$)	(5) 119 SCZ	(5) Ho *et al.* (2007)
COMT			
(1) Val158Met	(1) FL, GM, WM, CSF: n.s. ($F = < 1.43$, df = 2.148, $P > 0.24$)	(1) 100 SCZ, 49 C	(1) Ho *et al.* (2005)
(2) Polymorphisms	(2) Val COMT: ACC (bilateral) $p = 0.007$, MTG (R) = 0.016, Amygdala-uncus (L) $p = 0.01$, thalamus (left) $p = 0.014$	(2) 30 SCZ, 104 C	(2) Ohnishi *et al.* (2006)
(3) Val allele, PRODH (rs20086720)	(3)↑GM density in right superior temporal gyrus (AA genotype of COMT), $p = 0.039$,↓WM density bilateral frontal lobes (GG genotype of PRODH), $p < 0.05$, COMT val allele epistasis with PRODH alleles: ↑WM left anterior lobe	(3) 51 SCZ (M)	(3) Zinkstok *et al.* (2007)

(*continued*)

TABLE VI (*continued*)

Genes (alleles)	Association of genes/alleles with brain structures	Sample size	Study ID
DISC1			
(1) Rare DISC/ TRAX haplotypes	(1) HEP1 allele: ↓focal GM density sup + inf frontal gyri ($p < 0.05$) ($p < 8.61$ x10–31 and $P < 1.21 \times 10$–(8) ↓HC volume ($\chi^2 = 3.42$, $p = 0.06$, trend), HEP2/HEP3 allele: n.s. HC volume change ($p = 0.58$), AATG Haplotype:↓focal GM density in superior + middle frontal gyri and superior temporal gyrus and superior parietal cortex	(1) 6 MZ + 1 DZ twins pairs cc for SCZ, 28 MZ and 31 DZ C twin pairs	(1) Cannon *et al.* (2005)
(2) Serine/ Cysteine allele	(2) ↓HC gray matter volume (Serine allele) (d = 0.36, $p < 0.05$)	(2) 86 SCZ (Serine), 72 C (Cysteine)	(2) Callicott *et al.* (2005)
APOE ε4	(1) HC: $t = 0.23$, df = 33, $p = 0.82$, TL volume: $t = 0.27$, df = 33, $p = 0.79$ (2) HC: n.s. ($T = 1.9$, df = 19, $p < 0.10$), Cerebral Asymmetry: n.s. (T1.9, df = 19, $P < 0.10$)	(1) 40 SCZ COS, 57 C (2) 21 SCZ	(1) Fernandez *et al.* (1999) (2) Hata *et al.* (2002)
IL 1	(1) GM volume: Frontal: $F = 6.06$, df = 1, $p < 0.02$, Temporal: $F = 11.62$, df = 1, $p = 0.001$, WM volume whole brain: $F = 5.54$, df = 1, $p < 0.03$ (2) ↑ Ventricles: Left: $t = 3.504$, df-21, $p = 0.002$, Right: $t = 2.784$, df = 21, $p = 0.01$, HC: $t < 1.6$, df = 21, p = n.s., DLPFC GM: $T < 1.5$, df = 21, p = n.s. ($p = 0.06$)	(1) 44 SCZ, 48 C (2) 23 SCZ, 45 C	(1) Meisenzahl *et al.* (2001) (2) Papiol *et al.* (2005)

Summary of published studies by August 2009 (PubMed and Medline citations) on genes for brain structural phenotypes including study results from patients with schizophrenia (SCZ) compared to healthy controls (C) in most studies. The studies report on the brain structure that is investigated, the sample size, the results of the intereaction between genes (alleles/SNPs, single nucleotide polymorphisms) and brain morphological changes (significant or nonsignificant association (n.s.), a decrease (↓) or increase (↑) in brain volume/density in interaction with a gene or an allele or a SNP).

SCZ, schizophrenia; SCZ COS, schizophrenia childhood onset (before age 12); C, controls; cc, concordant and disc, discordant; HC, hippocampus; GM volume, gray matter volume; WM volume, white matter volume; CSF volume, cerebrospinal fluid spaces volume; DLPFC GM, dorsolateral prefrontal cortex gray matter; ACC, anterior cingulate cortex; MTG, middle temporal gyrus; TL volume, Temporal lobe volume; BDNF, brain derived neurotrophic factor; COMT, catechol-*o*-methyl transferase; RGS4, regulators of G protein signalling; NRG1, neuroglin 1; PP, prion protein gene; NT3 gene, neurotrophin 3 gene; APOE, apolipoprotein E; PRNP, prion protein; DISC1, disrupted in schizophrenia 1; IL-1RN, interleukin -1 receptor antagonist; TNF, tumor necrosis factor; NOTCH4 gene; PCM1 gene; PLXN B3, plexin B3 gene; GAD1 gene; ph, polymorphism; L, left; R, right; n.s., not significant; trios, both parents and a child; vol, volume; ant, anterior; sup, superior; inf, inferior.

TABLE VII

GENES FOR STRUCTURAL BRAIN PHENOTYPES IN PATIENTS WITH SCHIZOPHRENIA COMPARED TO HEALTHY CONTROLS, GENES STUDIED IN SINGLE STUDIES

Genes (alleles)	Association of genes/alleles with brain structures	Sample size	Study ID
NT gene (A3 allele)	HC volume: Mean Left HC: $t = 1.87$, df = 10, $p = 0.091$, Mean Right HC: $t = 2.72$, df = 10, $p = 0.022$	12 SCZ	Kunugi *et al.* (1999)
TNF-RII (homozygous allele 1)	Allele 1: ↑ ventricle volume, ↓ Frontal lobe volume	140 probands with SCZ, 197 parents of probands, 46 C	Wassink *et al.* (2000)
PRNP gene (Met codon 129 allele)	WM volume: $F = 5.285$, df = 1/84, $p = 0/024$, CSF volume: $F = 4.400$, df = 1/84, $p = 0.039$, GM volume: $F = 1.843$, df = 1/84, $p = 0.178$ (n.s.)	43 SCZ and 47 C	Rujescu *et al.* (2002)
NOTCH4 $(CTG)_\eta$	Frontal lobe (GM, WM, CSF): n.s. ($\chi = 7.09$, df = 6, $p = 0.31$)	158 SCZ, 48 C	Wassink *et al.* (2003)
GAD1	GM volume: ↓ GM volume in 6 out of 10 SNPs ($p = 0.04$–0.003)	72 SCZ COS, 25 C	Addington *et al.* (2005)
RGS4 (non G allele)	DLPFC: ↓ Left DLPFC GM volume, SNP4 ($p = 0.0072$), SNP18 ($p = 0.012$), ↓ Right DLFPC GM volume, SNP1 ($p = 0.0024$), SNP4 ($p = 0.012$), SNP7 ($p = 0.022$)	30 SCZ, 27 C	Prasad *et al.* (2005)
PCM1 (allele + or − on chromosome 8)	Gray matter bilateral orbitofrontal: ↓ GM volume, $F = 4.6$, $p < 0.05$ Gray matter in temporal lobe, HC, inferior temporal cortex: ↓ GM, $F = 12.7$, $p < 0.01$	14 SCZ (allele +), 14 C 13 SCZ (allele −), 14 C	Gurling *et al.* (2006)
NRG1 (risk allele 420M9–1395)	GM trajectories, volumes: ↑ ($p = 0.001$) WM trajectories, volumes: ↑ ($p = 0.04$)	78 SCZ COS, 165 C	Addington *et al.* (2007)
PLXNB3 (haplotype A+)	WM: $F = 4.275$, df = 1/85, $p = 0.042$, GM: $F = 2.362$, df = 1/85, $p = 0.128$, CSF: $F = 5.302$, df = 1/85, $p = 0.024$	48 SCZ, 48 C	Rujescu *et al.* (2007)

Summary of published studies by August 2009 (PubMed and Medline citations) on genes for brain structural phenotypes including study results from patients with schizophrenia (SCZ) compared to healthy controls (C) in most studies. The studies report on the brain structure that is investigated, the sample size, the results of the intereaction between genes (alleles/SNPs, single nucleotide polymorphisms) and brain morphological changes (significant or nonsignificant association (n.s.), a decrease (↓) or increase (↑) in brain volume/density in interaction with a gene or an allele or a SNP).

SCZ, schizophrenia; SCZ COS, schizophrenia childhood onset (before age 12); C, controls; cc, concordant and disc, discordant; HC, hippocampus; GM volume, gray matter volume; WM volume, white matter volume; CSF volume, cerebrospinal fluid spaces volume; DLPFC GM, dorsolateral prefrontal cortex gray matter; ACC, anterior cingulate cortex; MTG, middle temporal gyrus; TL volume, Temporal lobe volume; BDNF, brain derived neurotrophic factor; COMT, catechol-*o*-methyl transferase; RGS4, regulators of G protein signalling; NRG1, neuroglin 1; PP, prion protein gene; NT3 gene, neurotrophin 3 gene; APOE, apolipoprotein E; PRNP, prion protein; DISC1, disrupted in schizophrenia 1; IL-1RN, interleukin -1 receptor antagonist; TNF, tumor necrosis factor; NOTCH4 gene; PCM1 gene; PLXN B3, plexin B3 gene; GAD1 gene; ph, polymorphism; L, left; R, right; n.s., not significant; trios, both parents and a child; vol, volume; ant, anterior; sup, superior; inf, inferior.

frontal lobe. Ho *et al.* (2007) found an association between Met allele and a decrease in frontal gray matter and with an increase in lateral ventricles and in CSF sulcal volume and an increase in CSF volume in the frontal and temporal lobe, but not in the parietal and occipital lobe. Agartz *et al.* (2006) found a significant association of the 11757 G/C polymorphism with larger frontal gray matter and a 270C/T ph association with caudate volume and gray matter volume in caudate and no association with hippocampus, nor with brain volumes in controls except for the association between the −633 T/A polymorphisms and putamen. This *COMT* gene is the second most investigated gene and is located in a region of chromosome 22q11, which is deleted in patients with the velo-cardio-facial syndrome. In 25% of these patients, schizophrenia is diagnosed as well. In a recent metanalyses the *COMT* gene has not been strongly associated with schizophrenia (Williams *et al.*, 2007). *DISC1* gene is one of the genes found to be associated with several psychiatric disorders, such as schizophrenia, bipolar disorder, and depression (Hennah *et al.*, 2006). This gene was studied in a large Scottish family study, where a translocation was found to be inherited with different psychiatric disorders (Porteous, 2008). In persons with schizophrenia, there was a disruption of a gene of unknown function on one of the chromosomal breakpoints, and was therefore named disrupted in schizophrenia 1. This gene and the enzymes it codes for are of influence to neuronal migration and memory. In the study by Cannon *et al.* (2005), logistic regression was used to evaluate associations of DISC1 and TRAX haplotypes with diagnostic status, entering age, sex, substance use disorder, and parental social class as covariates and controlling for the dependency of observations within twin pairs using generalized estimation equations. The data of this study show that one of the three haplotypes (HEP 1) was associated with schizophrenia. The HEP1 haplotype (i.e., TCG) defines a 3-SNP segment located near the translocation break point of DISC1. The HEP1 haplotype was more often associated with individuals with schizophrenia (61.0%; $\chi^2 = 6.8$ [$p = 0.02$]; OR = 2.6; 95% CI = 1.2–5.6) when a recessive mode of transmission was assumed, but not among their nonschizophrenic co-twins (49.1%; $\chi^2 = 0.8$ [$p = 0.38$]). This haplotype was also significantly associated with small, focal reductions of gray matter in the superior and inferior frontal gyri. The HEP1 haplotype also shows evidence of an association with hippocampal volume reduction ($\chi^2 = 3.42$, $p = 0.06$). The regions affected by DISC1 and TRAX haplotypes identified in these maps have previously been shown to be highly heritable in normal twins (Thompson *et al.*, 2001). These regions are also sensitive to the degree of genetic relationship to an affected individual in twins discordant for schizophrenia (Cannon *et al.*, 2002; Van Erp *et al.*, 2004).

The APOE ε4 allele is a major risk allele for late-onset Alzheimer's disease (Mahaney *et al.*, 1993). The study by Hata *et al.* (2002) shows similar findings to studies carried out in patients with Alzheimer's disease (Soininen and Riekkinen, 1996; Tohgi *et al.*, 1997) in that there is a decrease in hippocampal volume of the ε4 allele carriers compared to noncarriers. The IL-RII gene has been studied twice until now, one showing an association between a high activity allele in the interleukin-1b (IL-1b) gene and orbitofrontotemporal volume reductions of gray and white matter in schizophrenia subjects but not in controls (Meisenzahl *et al.*, 2001) and the other study showing variability at the *IL-1RN* gene contributing to the ventricular volumetric variation in patients with schizophrenia. The PP gene, with a Val/Met polymorphism at position 129 in the prion protein gene (*PRNP*), is known as a risk factor for Creutzfeldt-Jakob disease, which is accompanied by psychosis in the early stages. It is unclear whether this *PP* gene is also associated with schizophrenia (Martorell *et al.*, 2007), but in the study by Rujescu *et al.* (2002), there was a decrease in white matter volume and larger CSF volume in both the controls as well as the patients with the Met homozygotes. The other genes are studied in just one study. These studies have all been analyzed in detail in previous reviews (Pearlson and Calhoun, 2007; Van Haren *et al.*, 2008).

While the reviewed studies suggest high heritability rates for some brain structures, limitations on the neuroimaging, twin studies and NHP studies make the interpretation and the generalizability of their findings difficult.

IV. Limitations and Clinical Relevance of the Studies

A. Limitations of Neuroimaging Studies

A number of limitations in imaging studies need to be acknowledged, such as the validity and reliability of measures (e.g., using measurements that rely on a representative slice vs. whole volume or do not correct for total brain volume). Measurements of brain volumes reveal differences between affected subjects and healthy controls, but the magnitude of these differences is moderate and there is always a substantial overlap in the distributions of the comparison groups. However, looking at the heritability of brain volumes and shapes in healthy twin studies shows high heritability rates of these brain measurements. One solution to this problem might be to shape analysis to quantify subtle differences in brain structures, in order to enhance the identification of structural brain abnormalities that might be associated with neurological and psychiatric disorders such as schizophrenia. Another limitation of the reported studies is the sample size, which is a major issue since imaging techniques and genetic studies require considerable statistical power. Although the gross differences in size or symmetry of brain structures might be quantified, the individual cells and cell layers cannot yet be visualized. This means that, although the volume and shape of brain structures may be determined, the underlying cause of any differences cannot. Another limitation in imaging studies is the method applied to measure the volumes of brain structures. Some studies use the ROI analyses, where specific brain areas are manually traced. This method does not accommodate changes at sites distant to the ROI, and thus may be inherently biased in spatial terms. Investigating covariation between structures is important, for it may suggest that different brain structures are being influenced by the same gene or sets of genes. This approach may also be less sensitive to small variations in brain structure or tissue type distribution than more modern modeling methods (Good *et al.*, 2001a, 2002). There is also a wide intra- and inter-rater variability, and ROI measurement errors are likely to be greater for smaller structures with complex architectures, particularly if anatomical borders are defined by arbitrary criteria. Other studies used the VBM method to measure volumes of brain structures, a fully automated whole brain technique. The VBM method avoids many of the constraints of ROI analyses, but it incorporates a series of preprocessing steps that may complicate the simple interpretation of regional changes of gray matter loss. One new method used in the heritability estimates is the elastic deformation procedure, used by Thompson *et al.* (2001), in which researchers modeled the genetic influences on neuroanatomic variation between healthy MZ and DZ twin pairs at each voxel on the surface of the cortex. Findings from these analyses showed a significant influence of the genetic factors on the gray matter density in subregions of prefrontal and temporal lobes, particularly in Broca's and Wernicke's language areas. Other areas within these structures were found less heritable, suggesting that heritability estimates based on gross lobular regions of interest blur more subtle local differences.

B. Limitations of Twin Studies

Twin studies cannot solve the problem of sources of variance, such as the gene–environment interactions, nonrandom assortative mating, and the possibility of differential interactions of environment in MZ and DZ twin pairs and the chorion type in MZ twins. Reliable detection of differences in correlations may therefore require hundreds of twin pairs. Interpretation of all the data in the reported studies is based on three assumptions in the classical twin designs: (1) MZ twins share 100% and DZ twins share on average 50% of their polymorphic genetic material and current methods adequately identify zygosity, (2) the environment shared among MZ and DZ twins is similar, and (3) twins are similar to singletons such that findings in twins can be generalized to nontwin populations. Conflicting results exist on the first assumption in schizophrenia, specifically that MZ twins discordant for schizophrenia share the same genetic material. Some studies (Guidry *et al.*, 1999; Petronis and Kennedy, 1995; Petronis *et al.*, 2003; Singh *et al.*, 2002) suggest that twins discordant for schizophrenia share less of their polymorphic material than concordant and healthy twins in contrast to other studies claiming there is no empirical evidence for this (Nguyen *et al.*, 2003; Tsujita *et al.*, 1998; Vincent *et al.*, 1998). The second assumption, that twins share the same environmental factors, has been criticized because of the physical maternal stress, which may be higher for twins then for singletons. There may also be more competition between the twins regarding nutrition. Although classic twin studies attempt to estimate unique environmental effects (events occurring to one twin but not to the other), as well as genetic (heritability) and environmental effects, one limitation of the model is that the gene–environment correlation and interaction is not discernible as a distinct effect. Another limitation is that there is no opportunity in the classic twin design to consider both shared environment and nonadditive genetic effects simultaneously. This limitation can be addressed by including additional siblings to the design, which was done by some studies in the healthy twin studies (Tables I and II). None of the studies reported in this review indicate the impact of possible prenatal and perinatal obstetric complications, except for the study by Van Erp *et al.* (2004) in which the obstetric records of half of the studied twin pairs did not differ between discordant MZ and DZ twin pairs, and the study by Wallace *et al.* (2006) in which data were corrected for the 29 weeks postgestational weight of the individuals included in the study. The third assumption on the generalizability of twin data to singletons is not possible according to some, because twins are not a random sample of the population and they differ in their developmental environment; in this sense they are not representative (Kempthorne and Osborne, 1961). Other studies claim that twins differ very little from their non-twin siblings and that measured studies on personality and intelligence of twins suggests that they have scores on these traits very similar to those of nontwins (Deary *et al.*, 2005). Only one study, by Hulshoff Pol *et al.* (2002), addresses this question and suggests that the data from twin studies can be generalized to singletons, when correcting for the intracranial volume for the second born twin compared to the first born, since second born twins have lower intracranial volume. All other volumes are comparable as well, suggesting that twin studies provide reliable estimates of heritability and can be generalized to singletons.

C. Limitations of Nonhuman Primate Studies

Due to the close anatomical, physiological, biochemical, and genetic similarity to humans, NHPs play an important role as model in many aspects of research, specifically for genetic analysis of individual variation in brain structure. The most used species are the baboons, belonging to the old-world monkeys the rhesus macaques, cynomolgus macaques, and vervet monkeys, and being very closely related to the humans. Other primate species that are more similar to humans are the chimpanzees and gorillas. These species would represent a better model for the human neuroanatomy, but they present only minimal opportunities for intraspecies genetic analyses due to the lack of extensive multigeneration pedigrees (Rogers *et al.*, 2007). Due to the use of large colonies, the analysis of complex phenotypes in the old world monkey can help us understand more about the genetic basis of individual variation in important traits.

One of the limitations of the NHP studies is that the heritability estimates in monkeys are specific to the population and circumstances in which they are assessed. Genetically diverse populations in homogeneous environments demonstrate larger heritability than do inbred populations in diverse environments (Falconer and Mackay, 1996).

Very few parents in the monkey colonies that make up study cohorts share a common mother or father, and extended family pedigrees cannot be determined from monkey breeding colony records. The generality of the NHP studies should therefore be tested in studies of other populations (Lyons *et al.*, 2001).

It may be clear that the human brain is more than simply a large monkey brain. On the one hand, the monkey and the human brains share a common plan of organization, but on the other hand, striking species differences and specific regional differences exist. In the course of evolution, besides an increase in mean size, the brain of the newborn human is exposed to the physical, social, and cultural environmental factors for an extended period compared to the monkey, resulting in "cultural imprints," such as learning the spoken and written language (Deheane *et al.*, 2005).

D. Clinical Relevance of the Heritability Estimations

Study of the influence of genetic factors on brain structures will help us to understand normal development and age, gender, and other variable associated changes, and will moreover help us understand the meaning of morphological changes found in schizophrenia.

One of the functional relevancies of the heritability of brain volumes is demonstrated best by the association between brain volumes and intelligence. In a study by Posthuma *et al.* (2002b), the association between brain volumes and intelligence was found to be of genetic origin. In two other studies (Thompson *et al.*, 2001; Thompson *et al.*, 2002), the association between frontal gray matter volume and intelligence was shown to be due to genetic factors. In a recent study by Hulshoff Pol *et al.* (2006), an association between intelligence and gray matter of the frontal and occipital lobe, the parahippocampus and the connecting white matter was found. While these findings suggest a common set of genes and intelligence, a study by Carmelli *et al.* (1998), shows that in elderly twins, the association between frontotemporal brain volumes, and executive function were found to be associated with common environmental influences shared by twins from the same family. Whether there is a gene–environment interaction, for example, persons with high level of cognitive functioning leading to select an environment that also increases brain size, or the possibility of overlapping sets of genes or common environmental influences causing variation in two distinct phenotypes, the cause of the findings between intelligence, and these brain structures, remains unknown. The heritability estimates are necessary to estimate the degree to which genetic factors and environmental factors have influence on the phenotype. Estimation of the impact of environmental factors is much more complicated and generally the effect size of environmental factors are considered to be the remaining part of the total heritability of a disease once the heritability due to genetic factors is known. In addition to the heritability measurements, genes involved in brain structures that are under the influence of genetic factors would be more easily investigated than looking at genes for brain structures in patients with schizophrenia, without knowing the origin (e.g., genetic vs. environmental, a gene–environment interaction or disease related) of these brain changes. This is where the studies in NHPs might be useful, specifically since the impact of the environmental factors can be controlled and questions regarding the evolution of different brain parts and of human cognition only can be addressed adequately if the differences and similarities in the underlying neural circuitry of humans and all extant apes can be identified. Some data from older studies have been used by investigators to address these issues, but problems regarding the methodology used in the measurements, the incomplete representation of hominoid species and the small samples included makes the interpretation of these data difficult.

E. Implications of These of Brain Volumes as Endophenotypes for Genetics of Psychiatric Disorders

Despite the limitations of the twin studies and neuroimaging studies, it is fair to say that these studies only represent a subset of the possible genetic and environmental relationship and that the heritability estimates from these studies represent a first step in understanding the genetics of brain structures.

Using the brain structures with high heritability scores as endophenotypes and looking at the gene action for these endophenotypes, might solve problems in neuroimaging and genetic studies. In this study we found that global structures, with high heritability rates, such as total brain volume, hemispheres, cerebrum, cerebellum, gray and white matter, the four lobes (frontal, temporal, occipital and parietal), and corpus callosum seem to be more useful as endophenotype markers than specific structures with lower heritability rates. These structures are indeed formed earlier in development and the high heritability rates are suggestive of more genetic influences on early developed structures. Lower heritability rates for the specific structures might be due to lower sample sizes found in most of the studies, due to differences in methods measuring the volumes or due to difference in age in participating persons. The studies in NHPs show the same results on the heritability of gross brain structures, although there are few studies on the heritability of brain structures in NHPs. Looking at the genes for brain structures in healthy persons and persons with schizophrenia shows us that gene approach studies do not show a causal pathway for the brain morphological changes in patients with schizophrenia, and no susceptibility gene with a high effect size has yet been found for schizophrenia.

To summarize, the findings of the reviewed studies suggest that the overall brain volume is highly heritable, comparable with twin based estimates for the most highly genetically determined human traits, including fingerprint ridge count ($h^2 = 0.98$) (Spence *et al.*, 1977), height ($h^2 = 0.66$) (Lettre, 2009; Silventoinen, 2003), and systolic blood pressure ($h^2 = 0.57$) (Williams *et al.*, 1990). The structures that are formed earlier in development (Rubenstein and Rakic, 1999; Rubenstein *et al.*, 1999) and located deeper in the brain may be more genetically influenced thus, they show higher heritability scores than the structures which are formed later in development (Rakic, 1988) or are located on surface areas (Nadarajah and Parnavelas, 2002; Rakic, 1972). These structures are also under more environmental influence. Heritability studies on the genetic control of the human brain structure and how much individual genotype accounts for the wide variations among individual brains using neuranatomically defined endophenotypes will make it easier to look for the actions of genes on brain structures. However, a critical look at the genetic and or disease related changes in the brains of patients with complex disorders such as schizophrenia is necessary, as is suggested by Cannon *et al.* (2002). Further research on the influences of genetic, common, and unique environmental factors is necessary with advanced quantitative genetic methods. Genotyping studies in search of genes for brain structures in healthy subjects and for brain morphological changes in diseased need to be replicated; however, until now individual studies have been generally underpowered in replicating the small number of promising susceptibility genes (Ross *et al.*, 2006) with small effect sizes (OR < 1.5). Looking for an association between brain morphological changes with genes, in particular the search for rare variants for overall genetic risk, with a more advanced method like the Genome Wide Association studies in combination with brain morphological changes, could bring new findings in causal pathways wherein genes interact with brain morphology. This may also shed light on understanding the relationship between schizophrenia and other major psychiatric disorders.

Acknowledgments

This research was supported by the Netherlands Organization for Scientific Research (NWO) under project number 017.002.048. We also thank A. Jasinska, Ph.D., UCLA/Los Angeles/USA, for her advices.

References

Addington, A. M., Gornick, M., Duckworth, J., Sporn, A., Gogtay, N., Bobb, A., Greenstein, D., Lenane, M., Gochman, P., Baker, N., Balkissoon, R., Vakkalanka, R. K., *et al.* (2005). GAD1 (2q31.1), which encodes glutamic acid decarboxylase (GAD67), is associated with childhood-onset schizophrenia and cortical gray matter volume loss. *Mol. Psychiatry* **10**(6), 581–588.

Addington, A. M., Gornick, M. C., Shaw, P., Seal, J., Gogtay, N., Greenstein, D., Clasen, L., Coffey, M., Gochman, P., Long, R., and Rapoport, J. L. (2007). Neuregulin 1 (8p12) and childhood-onset schizophrenia: Susceptibility haplotypes for diagnosis and brain developmental trajectories. *Mol. Psychiatry* **12**(2), 195–205.

Agartz, I., Sedvall, G. C., Terenius, L., Kulle, B., Frigessi, A., Hall, H., and Jönsson, E. G. (2006). BDNF gene variants and brain morphology in schizophrenia. *Am. J. Med. Genet. B Neuropsychiatr. Genet* **141B**(5), 513–523.

Alexander, M. P., and Annett, M. (1996). Crossed aphasia and related anomalies of cerebral organization: Case reports and a genetic hypothesis. *Brain Lang.* **55**(2), 213–239. Review.

Allen, J. S., Damasio, H., and Grabowski, T. J. (2002). Normal neuroanatomical variation in the human brain: An MRI-volumetric study. *Am. J. Phys. Anthropol.* **118**(4), 341–358.

Angelucci, F., Brenè, S., and Mathé, A. A. (2005). BDNF in schizophrenia, depression and corresponding animal models. *Mol. Psychiatry* **10**(4), 345–352. Review.

Annett, M. (1974). Handedness in the children of two left-handed parents. *Br. J. Psychol.* **65**(1), 129–131.

Ashburner, J., and Friston, K. J. (2000). Voxel-based morphometry: The methods. *Neuroimage* **11** (6 Pt 1), 805–821. Review.

Atwood, L. D., Wolf, P. A., Heard-Costa, N. L., Massaro, J. M., Beiser, A., D'Agostino, R. B., and DeCarli, C. (2004). Genetic variation in white matter hyperintensity volume in the Framingham Study. *Stroke* **35**(7), 1609–1613.

Baaré, W. F., Hulshoff Pol, H. E., Boomsma, D. I., Posthuma, D., de Geus, E. J., Schnack, H. G., van Haren, N. E., van Oel, C. J., and Kahn, R. S. (2001a). Quantitative genetic modeling of variation in human brain morphology. *Cereb Cortex* **11**(9), 816–824.

Baaré, W. F., van Oel, C. J., Hulshoff Pol, H. E., Schnack, H. G., Durston, S., Sitskoorn, M. M., and Kahn, R. S. (2001b). Volumes of brain structures in twins discordant for schizophrenia. *Arch. Gen. Psychiatry* **58**(1), 33–40.

Bartley, A. J., Jones, D. W., and Weinberger, D. R. (1997). Genetic variability of human brain size and cortical gyral patterns. *Brain* **120**((Pt 2), 257–269.

Bentall, R. P., Read, J. E., and Mosher, L. R. (2004). Models of Madness: Psychological, Social and Biological Approaches to Schizophrenia Brunner-Routledge, Philadelphia. ISBN 1-58391-906-6.

Biondi, A., Nogueira, H., Dormont, D., Duyme, M., Hasboun, D., Zouaoui, A., Chantôme, M., and Marsault, C. (1998). Are the brains of monozygotic twins similar? A three-dimensional MR study. *AJNR Am. J. Neuroradiol.* **19**(7), 1361–1367.

Blangero, J., Williams, J. T., and Almasy, L. (2003). Novel family-based approaches to genetic risk in thrombosis. *J. Thromb. Haemost* **1**(7), 1391–1397.

Boyle, M. (2002). Schizophrenia: A Scientific Delusion? Routledge, New York. ISBN 0-415-22718-6.

Boos, H. B., Aleman, A., Cahn, W., Hulshoff Pol, H., and Kahn, R. S. (2007). Brain volumes in relatives of patients with schizophrenia: A meta-analysis. *Arch. Gen. Psychiatry* **64**(3), 297–304.

Boomsma, D., Busjahn, A., and Peltonen, L. (2002). Classical twin studies and beyond. *Nat. Rev. Genet.* **3**(11), 872–882. Review.

Brans, R. G., van Haren, N. E., van Baal, G. C., van Baal, G. C., Schnack, H. G., Kahn, R. S., and Hulshoff Pol, H. E. (2008). Heritability of changes in brain volume over time in twin pairs discordant for schizophrenia. *Arch. Gen. Psychiatry* **65**(11), 1259–1268.

Bremner, J. D., Krystal, J. H., Southwick, S. M., and Charney, D. S. (1995). Functional neuroanatomical correlates of the effects of stress on memory. *J. Trauma Stress* **8**(4), 527–553. Review.

Bremner, J. D., Narayan, M., Anderson ER Staib, L. H., Miller, H. L., and Charney, D. S. (2000). Hippocampal volume reduction in major depression. *Am. J. Psychiatry* **157**(1), 115–118.

Bruckner, C. M., and Slanger, W. D. (1986). Symmetric differences squared and analysis of variance procedures for estimating genetic and environmental variances and covariances for beef cattle weaning weight: (a) Comparison via simulation (b) Estimates from a data set. *J. Anim. Sci.* **63**(6), 1779–1793.

Buckholtz, J. W., Meyer-Lindenberg, A., Honea, R. A., Straub, R. E., Pezawas, L., Egan, M. F., Vakkalanka, R., Kolachana, B., Verchinski, B. A., Sust, S., Mattay, V. S., Weinberger, D. R., *et al.* (2007). Allelic variation in RGS4 impacts functional and structural connectivity in the human brain. *J. Neurosci.* **27**(7), 1584–1593.

Bueller, J. A., Aftab, M., Sen, S., Gomez-Hassan, D., Burmeister, M., and Zubieta, J. K. (2006). BDNF Val66Met allele is associated with reduced hippocampal volume in healthy subjects. *Biol. Psychiatry* **59**(9), 812–815.

Cahn, W., Hulshoff Pol, H. E., Bongers, M., Schnack, H. G., Mandl, R. C., Van Haren, N. E., Durston, S., Koning, H., Van Der Linden, J. A., and Kahn, R. S. (2001). Brain morphology in antipsychotic-naïve schizophrenia: A study of multiple brain structures. *Br. J. Psychiatry.* **43** (Suppl.), s66–s72.

Callicott, J. H., Straub, R. E., Pezawas, L., Egan, M. F., Mattay, V. S., Hariri, A. R., Verchinski, B. A., Meyer-Lindenberg, A., Balkissoon, R., Kolachana, B., Goldberg, T. E., and Weinberger, D. R. (2005). Variation in DISC1 affects hippocampal structure and function and increases risk for schizophrenia. *Proc. Natl Acad. Sci. USA* **102**(24), 8627–8632.

Cardno, A. G., and Gottesman, A. G. (2000). Twin studies of schizophrenia: From bow-and-arrow concordances to star wars Mx and functional genomics. *Am. J. Med. Genet* **97**(1), 12–17. Review.

Cardno, A. G., Thomas, K., and McGuffin, P. (2002). Clinical variables and genetic loading for schizophrenia: Analysis of published Danish adoption study data. *Schizophr. Bull.* **28**(3), 393–399.

Carmelli, D., DeCarli, C., Swan, G. E., Jack, L. M., Reed, T., Wolf, P. A., and Miller, B. L. (1998). Evidence for genetic variance in white matter hyperintensity volume in normal elderly male twins. *Stroke* **29**(6), 1177–1181.

Cannon, T. D., Kaprio, J., Lönnqvist, J., Huttunen, M., and Koskenvuo, M. (1998). The genetic epidemiology of schizophrenia in a Finnish twin cohort. A population-based modeling study. *Arch. Gen. Psychiatry* **55**(1), 67–74.

Cannon, T. D., Thompson, P. M., van Erp, T. G., Toga, A. W., Poutanen, V. P., Huttunen, M., Lonnqvist, J., Standerskjold-Nordenstam, C. G., Narr, K. L., Khaledy, M., Zoumalan, C. I., Dail, R., *et al.* (2002). Cortex mapping reveals regionally specific patterns of genetic and disease-specific gray-matter deficits in twins discordant for schizophrenia. *Proc. Natl Acad. Sci. USA* **99**(5), 3228–3233.

Cannon, T. D., van Erp, T. G., Bearden, C. E., Loewy, R., Thompson, P., Toga, A. W., Huttunen, M. O., Keshavan, M. S., Seidman, L. J., and Tsuang, M. T. (2003). Early and late neurodevelopmental influences in the prodrome to schizophrenia: Contributions of genes, environment, and their interactions. *Schizophr. Bull.* **29**(4), 653–669.

Cannon, T. D., Hennah, W., van Erp, T. G., Thompson, P. M., Lonnqvist, J., Huttunen, M., Gasperoni, T., Tuulio-Henriksson, A., Pirkola, T., Toga, A. W., Kaprio, J., Mazziotta, J., *et al.* (2005). Association of DISC1/TRAX haplotypes with schizophrenia, reduced prefrontal gray matter, and impaired short- and long-term memory. *Arch. Gen. Psychiatry* **62**(11), 1205–1213.

Cantalupo, C., and Hopkins, W. D. (2001). Asymmetric Broca's area in great apes. *Nature* **414**(6863), 505.

Carne, R. P., Vogrin, S., Litewka, L., and Cook, M. J. (2006). Cerebral cortex: An MRI-based study of volume and variance with age and sex. *J. Clin. Neurosci* **13**(1), 60–72.

Cheverud, J. M., Falk, D., Vannier, M., Moore, A. J., Helmkamp, R. C., and Vannier, M. (1990). Heritability of brain size and surface features in rhesus macaques (Macaca mulatta). *J. Hered* **81**(1), 51–57.

Cosgrove, K. P., Mazure, C. M., and Staley, J. K. (2007). Evolving knowledge of sex differences in brain structure, function, and chemistry. *Biol. Psychiatry* **62**(8), 847–855. Review.

Deary, I. J., Pattie, A., Wilson, V., and Whalley, L. J. (2005). The cognitive cost of being a twin: Two whole-population surveys. *Twin. Res. Hum. Genet* **8**(4), 376–383.

Dehaene, S., Duhamel, J.-R. D., Hauser, M. D., and Rizzolatti, G. (2005). Genes, brains, and culture: From monkey to human. In "Monkey Brain to Human Brain".

Deutch, A. Y. (1993). Prefrontal cortical dopamine systems and the elaboration of functional corticostriatal circuits: Implications for schizophrenia and Parkinson's disease. *J. Neural Transm. Gen. Sect.* **91**(2–3), 197–221. Review.

Diagnostic and Statistical Manual of Mental Disorders (DSM-IV-Tr) (2000). American Psychiatric Association (APA).

Driemeyer, J., Boyke, J., Gaser, C., Büchel, C., and May, A. (2008). Changes in gray matter induced by learning-revisited. *PLoS One* **3**(7), e2669.

Eckert, M. A., Leonard, C. M., Molloy, E. A., Blumenthal, J., Zijdenbos, A., and Giedd, J. N. (2002). The epigenesis of planum temporale asymmetry in twins. *Cereb. Cortex* **12**(7), 749–755.

Falconer, D. S., and Mackay, T. F. C. (1996). Introduction to Quantitative Genetics 4th Ed. Addison Wesley Longman, Harlow, Essex, UK.

Falconer, D. S., Gauld, I. K., Roberts, R. C., and Williams, D. A. (1981). The control of body size in mouse chimaeras. *Genet. Res* **38**(1), 25–46.

Fears, S. C., Melega, W. P., Service, S. K., Lee, C., Chen, K., Tu, Z., Jorgensen, M. J., Fairbanks, L. A., Cantor, R. M., Freimer, N. B., and Woods, R. P. (2009). Identifying heritable brain phenotypes in an extended pedigree of vervet monkeys. *J. Neurosci.* **29**(9), 2867–2875.

Fernandez, T., Yan, W. L., Hamburger, S., Rapoport, J. L., Saunders, A. M., Schapiro, M., Ginns, E. I., and Sidransky, E. (1999). Apolipoprotein E alleles in childhood-onset schizophrenia. *Am. J. Med. Genet* **88**(2), 211–213.

Gannon, P. J., Holloway, R. L., Broadfield, D. C., and Braun, A. R. (1998). Asymmetry of chimpanzee planum temporale: Humanlike pattern of Wernicke's brain language area homolog. *Science* **279** (5348), 220–222.

Geschwind, N., and Galaburda, A. M. (1985). Cerebral lateralization. Biological mechanisms, associations, and pathology: III. A hypothesis and a program for research. *Arch. Neurol.* **42**(7), 634–654.

Geschwind, D. H., Miller, B. L., DeCarli, C., and Carmelli, D. (2002). Heritability of lobar brain volumes in twins supports genetic models of cerebral laterality and handedness. *Proc. Natl. Acad. Sci. USA* **99**(5), 3176–3181.

Giedd, J. N. (2008). The teen brain: Insights from neuroimaging. *J. Adolesc. Health* **42**(4), 335–343. Review.

Good, C. D., Ashburner, J., and Frackowiak, R. S. (2001a). Computational neuroanatomy: New perspectives for neuroradiology. *Rev. Neurol. (Paris)* **157**(8–9 Pt 1), 797–806. Review.

Good, C. D., Johnsrude, I. S., Ashburner, J., Henson, R. N., Friston, K. J., and Frackowiak, R. S. (2001b). A voxel-based morphometric study of ageing in 465 normal adult human brains. *Neuroimage* **14**(1 Pt 1), 21–36.

Good, C. D., Scahill, R. I., Fox, N. C., Ashburner, J., Friston, K. J., Chan, D., Crum, W. R., Rossor, M. N., and Frackowiak, R. S. (2002). Automatic differentiation of anatomical patterns in the human brain: Validation with studies of degenerative dementias. *Neuroimage* **17**(1), 29–46.

Goldman, M. B., Torres, I. J., Keedy, S., Marlow-O'Connor, M., Beenken, B., and Pilla, R. (2007). Reduced anterior hippocampal formation volume in hyponatremic schizophrenic patients. *Hippocampus* **17**(7), 554–562.

Gottesman, I. I., and Gould, T. D. (2003). The endophenotype concept in psychiatry: Etymology and strategic intentions. *Am. J. Psychiatry* **160**(4), 636–645. Review.

Gould, E., and Tanapat, P. (1999). Stress and hippocampal neurogenesis. *Biol. Psychiatry* **46**(11), 1472–1479. Review.

Guidry, J., and Kent, T. A. (1999). New genetic hypothesis of schizophrenia. *Med. Hypotheses.* **52**(1), 69–75.

Gurling, H. M., Critchley, H., Datta, S. R., McQuillin, A., Blaveri, E., Thirumalai, S., Pimm, J., Krasucki, R., Kalsi, G., Quested, D., Lawrence, J., Bass, N., *et al.* (2006). Genetic association and brain morphology studies and the chromosome 8p22 pericentriolar material 1 (PCM1) gene in susceptibility to schizophrenia. *Arch. Gen. Psychiatry* **63**(8), 844–854. Erratum in: *Arch. Gen. Psychiatry* **64**(11), 1258.

Gurvits, T. V., Shenton, M. E., Hokama, H., Ohta, H., Lasko, N. B., Gilbertson, M. W., Orr, S. P., Kikinis, R., Jolesz, F. A., McCarley, R. W., and Pitman, R. K. (1996). Magnetic resonance imaging study of hippocampal volume in chronic, combat-related posttraumatic stress disorder. *Biol. Psychiatry* **40**(11), 1091–1099.

Hata, T., Kunugi, H., Nanko, S., Fukuda, R., and Kaminaga, T. (2002). Possible effect of the APOE epsilon 4 allele on the hippocampal volume and asymmetry in schizophrenia. *Am. J. Med. Genet* **114**(6), 641–642.

Hennah, W., Thomson, P., Peltonen, L., and Porteous, D. (2006). Genes and schizophrenia: Beyond schizophrenia—The role of DISC1 in major mental illness. *Schizophr. Bull.* **32**(3), 409–416. Review.

Ho, B. C., Wassink, T. H., O'Leary, D. S., Sheffield, V. C., and Andreasen, N. C. (2005). Catechol-*O*-methyl transferase Val158Met gene polymorphism in schizophrenia: Working memory, frontal lobe MRI morphology and frontal cerebral blood flow. *Mol. Psychiatry* **10**(3), 229, 287–298.

Ho, B. C., Milev, P., O'Leary, D. S., Librant, A., Andreasen, N. C., and Wassink, T. H. (2006). Cognitive and magnetic resonance imaging brain morphometric correlates of brain-derived neurotrophic factor Val66Met gene polymorphism in patients with schizophrenia and healthy volunteers. *Arch. Gen. Psychiatry* **63**(7), 731–740.

Ho, B. C., Andreasen, N. C., Dawson, J. D., and Wassink, T. H. (2007). Association between brain-derived neurotrophic factor Val66Met gene polymorphism and progressive brain volume changes in schizophrenia. *Am. J. Psychiatry* **164**(12), 1890–1899.

Hulshoff Pol, H. E., Posthuma, D., Baaré, W. F., De Geus, E. J., Schnack, H. G., van Haren, N. E., van Oel, C. J., Kahn, R. S., and Boomsma, D. I. (2002). Twin-singleton differences in brain structure using structural equation modelling. *Brain.* **125**(Pt 2), 384–390.

Hulshoff Pol, H. E., Brans, R. G., van Haren, N. E., Schnack, H. G., Langen, M., Baaré, W. F., van Oel, C. J., and Kahn, R. S. (2004). Gray and white matter volume abnormalities in monozygotic and same-gender dizygotic twins discordant for schizophrenia. *Biol. Psychiatry* **55**(2), 126–130.

Hulshoff Pol, H. E., Schnack, H. G., Posthuma, D., Mandl, R. C., Baaré, W. F., van Oel, C., van Haren, N. E., Collins, D. L., Evans, A. C., Amunts, K., Bürgel, U., Zilles, K., *et al.* (2006). Genetic contributions to human brain morphology and intelligence. *J. Neurosci.* **26**(40), 10235–10242.

Ilg, R., Wohlschläger, A. M., Gaser, C., Liebau, Y., Dauner, R., Wöller, A., Zimmer, C., Zihl, J., and Mühlau, M. (2008). Gray matter increase induced by practice correlates with task-specific activation: A combined functional and morphometric magnetic resonance imaging study. *J. Neurosci.* **28**(16), 4210–4215.

Kanaan, R., Barker, G., Brammer, M., Giampietro, V., Shergill, S., Woolley, J., Picchioni, M., Toulopoulou, T., and McGuire, P. (2009). White matter microstructure in schizophrenia: Effects of disorder, duration and medication. *Br. J. Psychiatry* **194**(3), 236–242.

Kappelman, J. (1993). The evolution of body mass and relative brain size in fossil hominoids. *J. Human Evol.* **30,** 243–276.

Kempthorne, O., and Osborne, R. H. (1961). The interpretation of twin data. *Am. J. Hum. Genet.* **13,** 320–339.

Kendell, R., and Jablensky, A. (2003). Distinguishing between the validity and utility of psychiatric diagnoses. *Am. J. Psychiatry* **160**(1), 4–12.

Kendler, K. S., and Diehl, S. R. (1993). The genetics of schizophrenia: A current, genetic-epidemiologic perspective. *Schizophr. Bull.* **19**(2), 261–285.

Koolschijn, P. C., van Haren, N. E., Hulshoff Pol, H. E., and Kahn, R. S. (2007). Hypothalamus volume in twin pairs discordant for schizophrenia. *Eur. Neuropsychopharmacol.* **18**(4), 312–315.

Kunugi, H., Hattori, M., Nanko, S., Fujii, K., Kato, T., and Nanko, S. (1999). Dinucleotide repeat polymorphism in the neurotrophin-3 gene and hippocampal volume in psychoses. *Schizophr. Res.* **37**(3), 271–273.

Lawrie, S. M., and Abukmeil, S. S. (1998). Brain abnormality in schizophrenia. A systematic and quantitative review of volumetric magnetic resonance imaging studies. *Br. J. Psychiatry* **172,** 110–120.

Lee, H., Devlin, J. T., Shakeshaft, C., Stewart, L. H., Brennan, A., Glensman, J., Pitcher, K., Crinion, J., Mechelli, A., Frackowiak, R. S., Green, D. W., and Price, C. J. (2007). Anatomical traces of vocabulary acquisition in the adolescent brain. *J. Neurosci.* **27**(5), 1184–1189.

Lettre, G. (2009). Genetic regulation of adult stature. *Curr. Opin. Pediatr.* **21**(4), 515–522.

Linscott, R. J., Allardyce, J., and van Os, J. (2009). Seeking verisimilitude in a class: A systematic review of evidence that the criterial clinical symptoms of schizophrenia are taxonic. *Schizophr. Bull.* [Epub ahead of print].

Lohmann, G., von Cramon, D. Y., and Steinmetz, H. (1999). Sulcal variability of twins. *Cereb. Cortex* **9**(7), 754–763.

Lyons, D. M., Yang, C., Sawyer-Glover, A. M., Moseley, M. E., and Schatzberg, A. F. (2001). Early life stress and inherited variation in monkey hippocampal volumes. *Arch. Gen. Psychiatry* **58**(12), 1145–1151.

Magariños, A. M., and McEwen, B. S. (1995). Stress-induced atrophy of apical dendrites of hippocampal CA3c neurons: Comparison of stressors. *Neuroscience.* **69**(1):83–88.

Mahaney, M. C., Williams-Blangero, S., Blangero, J., and Leland, M. M. (1993). Quantitative genetics of relative organ weight variation in captive baboons. *Hum. Biol* **65**(6), 991–1003.

Martorell, L., Valero, J., Mulet, B., Gutiérrez-Zotes, A., Cortés, M. J., Jariod, M., Pérez, M., Labad, A., and Vilella, E. (2007). M129V variation in the prison protein gene and psychotic disorders: Relationship to neuropsychological and psychopathological measures. *J. Psychiatr. Res.* **41**(10), 885–892.

McDonald, C., and Murphy, K. C. (2003). The new genetics of schizophrenia. *Psychiatry Clin. North Am.* **26**(1), 41–63. Review.

McGuffin, P., Famer, A. E., Gottesman, I. I., Murray, R. M., and Reveley, A. M. (1984). Twin concordance for operationally defined schizophrenia. Confirmation of familiality and heritability. *Arch. Gen. Psychiatry* **41**(6), 541–545.

McIntosh, A. M., Baig, B. J., Hall, J., Job, D., Whalley, H. C., Lymer, G. K., Moorhead, T. W., Owens, D. G., Miller, P., Porteous, D., Lawrie, S. M., and Johnstone, E. C. (2007). Relationship of catechol-*O*-methyltransferase variants to brain structure and function in a population at high risk of psychosis. *Biol. Psychiatry* **61**(10), 1127–1134.

McIntosh, A. M., Moorhead, T. W., Job, D., Lymer, G. K., Muñoz Maniega, S., McKirdy, J., Sussmann, J. E., Baig, B. J., Bastin, M. E., Porteous, D., Evans, K. L., Johnstone, E. C., *et al.* (2008). The effects of a neuregulin 1 variant on white matter density and integrity. *Mol. Psychiatry* **13**(11), 1054–1059.

McManus, I. C., and Bryden, M. P. (1991). Geschwind's theory of cerebral lateralization: Developing a formal, causal model. *Psychol. Bull.* **110**(2), 237–253. Review.

McNeil, T. F. (2000). Perinatal risk factors and schizophrenia: Selective review and methodological concerns. *Epidemiol. Rev.* **17**(1), 107–112. Review.

Meisenzahl, E. M., Rujescu, D., Kirner, A., Giegling, I., Kathmann, N., Leinsinger, G., Maag, K., Hegerl, U., Hahn, K., and Möller, H. J. (2001). Association of an interleukin-1beta genetic polymorphism with altered brain structure in patients with schizophrenia. *Am. J. Psychiatry* **158**(8), 1316–1319.

Murray, R. M. (1994). Neurodevelopmental schizophrenia: The rediscovery of dementia praecox. *Br. J. Psychiatry* **25**(Suppl.), 6–12. Review.

Nadarajah, B., and Parnavelas, J. G. (2002). Modes of neuronal migration in the developing cerebral cortex. *Nat Rev Neurosci.* **3**(6), 423–432. Review.

Narr, K. L., van Erp, T. G., Cannon, T. D., Woods, R. P., Thompson, P. M., Jang, S., Blanton, R., Poutanen, V. P., Huttunen, M., Lönnqvist, J., Standerksjöld-Nordenstam, C. G., Kaprio, J., *et al.* (2002). A twin study of genetic contributions to hippocampal morphology in schizophrenia. *Neurobiol. Dis.* **11**(1), 83–95.

Nguyen, G. H., Bouchard, J., Boselli MGTolstoi, L. G., Keith, L., Baldwin, C., Nguyen, N. C., Schultz, M., Herrera, V. L., and Smith, C. L. (2003). DNA stability and schizophrenia in twins. *Am. J. Med. Genet. B Neuropsychiatr. Genet.* **120B**(1), 1–10.

Ohnishi, T., Hashimoto, R., Mori, T., Nemoto, K., Moriguchi, Y., Iida, H., Noguchi, H., Nakabayashi, T., Hori, H., Ohmori, M., Tsukue, R., Anami, K., *et al.* (2006). The association between the Val158Met polymorphism of the catechol-*O*-methyl transferase gene and morphological abnormalities of the brain in chronic schizophrenia. *Brain* **129**(Pt 2), 399–410.

Okugawa, G., Nobuhara, K., Takase, K., and Kinoshita, T. (2007). Cerebellar posterior superior vermis and cognitive cluster scores in drug-naive patients with first-episode schizophrenia. *Neuropsychobiology* **56**(4), 216–219.

Oppenheim, J. S., Skerry, J. E., Tramo, M. J., and Gazzaniga, M. S. (1989). Magnetic resonance imaging morphology of the corpus callosum in monozygotic twins. *Ann. Neurol.* **26**(1), 100–104.

Papiol, S., Molina, V., Desco, M., Rosa, A., Reig, S., Gispert, J. D., Sanz, J., Palomo, T., and Fañanás, L. (2005). Ventricular enlargement in schizophrenia is associated with a genetic polymorphism at the interleukin-1 receptor antagonist gene. *Neuroimage* **27**(4), 1002–1006.

Pearlson, G. D., and Calhoun, V. (2007). Structural and functional magnetic resonance imaging in psychiatric disorders. *Can J. Psychiatry* **52**(3), 158–166. Review.

Pennington, B. F., Filipek, P. A., Lefly, D., Chhabildas, N., Kennedy, D. N., Simon, J. H., Filley, C. M., Galaburda, A., and DeFries, J. C. (2000). A twin MRI study of size variations in human brain. *J. Cogn. Neurosci.* **12**(1), 223–232.

Peper, J. S., Schnack, H. G., Brouwer, R. M., Van Baal, G. C., Pjetri, E., Székely, E., van Leeuwen, M., van den Berg, S. M., Collins, D. L., Evans, A. C., Boomsma, D. I., Kahn, R. S., *et al.* (2009). Heritability of regional and global brain structure at the onset of puberty: A magnetic resonance imaging study in 9-year-old twin pairs. *Hum. Brain Mapp.* **30**(7), 2184–2196.

Petronis, A., and Kennedy, J. L. (1995). Unstable genes: Unstable mind? *Am. J. Psychiatry* **152**(2), 164–172. Review.

Petronis, A., Gottesman, I. I., Kan, P., Kennedy, J. L., Basile, V. S., Paterson, A. D., and Popendikyte, V. (2003). Monozygotic twins exhibit numerous epigenetic differences: Clues to twin discordance? *Schizophr. Bull.* **29**(1), 169–178.

Pezawas, L., Verchinski, B. A., Mattay, V. S., Callicott, J. H., Kolachana, B. S., Straub, R. E., Egan, M. F., Meyer-Lindenberg, A., and Weinberger, D. R. (2004). The brain-derived neurotrophic factor val66met polymorphism and variation in human cortical morphology. *J. Neurosci.* **24**(45), 10099–10102.

Pfefferbaum, A., Sullivan, E. V., and Carmelli, D. (2004). Morphological changes in aging brain structures are differentially affected by time-linked environmental influences despite strong genetic stability. *Neurobiol. Aging* **25**(2), 175–183.

Plassman, B. L., Welsh-Bohmer, K. A., Bigler, E. D., Johnson, S. C., Anderson, C. V., Helms, M. J., Saunders, A. M., and Breitner, J. C. (1997). Apolipoprotein E epsilon 4 allele and hippocampal volume in twins with normal cognition. *Neurology* **48**(4), 985–989.

Posthuma, D., and Boomsma, D. I. (2000). A note on the statistical power in extended twin designs. *Behav. Genet.* **30**(2), 147–158.

Posthuma, D., de Geus, E. J., Neale, M. C., Hulshoff Pol, H. E., Baaré, W. F., Kahn, R. S., and Boomsma, D. (2000). Multivariate genetic analysis of brain structure in an extended twin design. *Behav. Genet.* **30**(4), 311–319.

Posthuma, D., Mulder, E. J., Boomsma, D. I., and de Geus, E. J. (2002a). Genetic analysis of IQ, processing speed and stimulus-response incongruency effects. *Biol. Psychol* **61**(1–2), 157–182.

Posthuma, D., De Geus, E. J., Baaré, W. F., Hulshoff Pol, H. E., Kahn, R. S., and Boomsma, D. I. (2002b). The association between brain volume and intelligence is of genetic origin. *Nat. Neurosci.* **5**(2), 83–84.

Posthuma, D., Baaré, W. F., Hulshoff Pol, H. E., Kahn, R. S., Boomsma, D. I., and De Geus, E. J. (2003). Genetic correlations between brain volumes and the WAIS-III dimensions of verbal comprehension, working memory, perceptual organization, and processing speed. *Twin Res.* **6**(2), 131–139.

Prasad, S., Chowdari, K. V., Wood, J., Bhatia, T., Deshpande, S. N., Nimgaonkar, V. L., and Thelma, B. K. (2005). Association analysis of NOTCH 4 polymorphisms with schizophrenia among two independent family based samples. *Am. J. Med. Genet. B Neuropsychiatr. Genet.* **131B**(1), 6–9.

Price, B. (1950). Primary biases in twin studies; a review of prenatal and natal difference-producing factors in monozygotic pairs. *Am. J. Hum. Genet.* **2**(4), 293–352.

Porteous, D. (2008). Genetic causality in schizophrenia and bipolar disorder: Out with the old and in with the new. *Curr. Opin Genet. Dev.* **18**(3), 229–234. Review.

Rais, M., Cahn, W., Van Haren, N., Schnack, H., Caspers, E., Hulshoff Pol, H., and Kahn, R. (2008). Excessive brain volume loss over time in cannabis-using first-episode schizophrenia patients. *Am. J. Psychiatry* **165**(4), 490–496.

Rakic, P. (1972). Mode of cell migration to the superficial layers of fetal monkey neocortex. *J. Comp. Neurol.* **145**(1), 61–83.

Rakic, P. (1988). Specification of cerebral cortical areas. *Science* **241**(4862), 170–176. Review.

Reveley, A. M., Reveley, M. A., Chitkara, B., *et al.* (1984). The genetic basis of cerebral ventricular volume. *Psychiatry Res.* **13**(3), 261–266.

Rijsdijk, F. V., van Haren, N. E., Picchioni, M. M., McDonald, C., Toulopoulou, T., Hulshoff Pol, H. E., Kahn, R. S., Murray, R., and Sham, P. C. (2005). Brain MRI abnormalities in schizophrenia: Same genes or same environment? *Psychol. Med.* **35**(10), 1399–1409.

Rogers, J., Kochunov, P., Lancaster, J., Shelledy, W., Glahn, D., Blangero, J., and Fox, P. (2007). Heritability of brain volume, surface area and shape: An MRI study in an extended pedigree of baboons. *Hum. Brain Mapp.* **28**(6), 576–583.

Ross, C. A., Margolis, R. L., Reading, S. A., Pletnikov, M., and Coyle, J. T. (2006). Neurobiology of schizophrenia. *Neuron* **52**(1), 139–153. Review.

Roth, G., and Dicke, U. (2005). Evolution of the brain and intelligence. *Trends Cogn. Sci.* **9**(5), 250–257.

Rubenstein, J. L., and Rakic, P. (1999). Genetic control of cortical development. *Cereb. Cortex* **9**(6), 521–523.

Rubenstein, J. L., Anderson, S., Shi, L., Miyashita-Lin, E., Bulfone, A., and Hevner, R. (1999). Genetic control of cortical regionalization and connectivity. *Cereb. Cortex* **9**(6), 524–532.

Rujescu, D., Meisenzahl, E. M., Giegling, I., Kirner, A., Leinsinger, G., Hegerl, U., Hahn, K., and Möller, H. J. (2002). Methionine homozygosity at codon 129 in the prion protein is associated with white matter reduction and enlargement of CSF compartments in healthy volunteers and schizophrenic patients. *Neuroimage* **15**(1), 200–206.

Rujescu, D., Meisenzahl, E. M., Krejcova, S., Giegling, I., Zetzsche, T., Reiser, M., Born, C. M., Möller, H. J., Veske, A., Gal, A., and Finckh, U. (2007). Plexin B3 is genetically associated with verbal performance and white matter volume in human brain. *Mol. Psychiatry* **12**(2), 190–194. 115.

Sapolsky, R. M., Uno, H., Rebert, C. S., and Finch, C. E. (1990). Hippocampal damage associated with prolonged glucocorticoid exposure in primates. *J. Neurosci.* **10**(9), 2897–2902.

Sánchez, M. M., Hearn, E. F., Do, D., Rilling, J. K., and Herndon, J. G. (1998). Differential rearing affects corpus callosum size and cognitive function of rhesus monkeys. *Brain Res* **812**(1–2), 38–49.

Sapolsky, R. M. (2000). Glucocorticoids and hippocampal atrophy in neuropsychiatric disorders. *Arch. Gen. Psychiatry* **57**(10), 925–935.

Scamvougeras, A., Kigar, D. L., Jones, D., Weinberger, D. R., and Witelson, S. F. (2003). Size of the human corpus callosum is genetically determined: An MRI study in mono and dizygotic twins. *Neurosci. Lett.* **338**(2), 91–94.

Schmitt, J. E., Lenroot, R. K., Wallace, G. L., Ordaz, S., Taylor, K. N., Kabani, N., Greenstein, D., Lerch, J. P., Kendler, K. S., Neale, M. C., and Giedd, J. N. (2008). Identification of genetically mediated cortical networks: A multivariate study of pediatric twins and siblings. *Cereb. Cortex* **18**(8), 1737–1747.

Semendeferi, K., Damasio, H., Frank, R., and Van Hoesen, G. W. (1997). The evolution of the frontal lobes: A volumetric analysis based on three-dimensional reconstructions of magnetic resonance scans of human and ape brains. *J. Hum. Evol.* **32**(4), 375–388.

Semendeferi, K., Lu, A., Schenker, N., Damasio, H., and Damasio, H. (2001). Humans and great apes share a large frontal cortex. *Nat. Neurosci.* **5**(3), 272–276.

Siegel, S. J., Ginsberg, S. D., Hof, P. R., Foote, S. L., Young, W. G., Kraemer, G. W., McKinney, W. T., and Morrison, J. H. (1993). Effects of social deprivation in prepubescent rhesus monkeys: Immunohistochemical analysis of the neurofilament protein triplet in the hippocampal formation. *Brain Res.* **619**(1–2), 299–305.

Silventoinen, K. (2003). Determinants of variation in adult body height. *J. Biosoc. Sci.* **35**(2), 263–285.

Singh, S. M., Murphy, B., and O'Reilly, R. (2002). Epigenetic contributors to the discordance of monozygotic twins. *Clin. Genet.* **62**(2), 97–103. Review.

Sheline, Y. I., Sanghavi, M., Mintun, M. A., and Gado, M. H. (1999). Depression duration but not age predicts hippocampal volume loss in medically healthy women with recurrent major depression. *J. Neurosci.* **19**(12), 5034–5043.

Soininen, H. S., and Riekkinen, P. J. Sr. (1996). Apolipoprotein E, memory and Alzheimer's disease. *Trends Neurosci.* **19**(6), 224–228.

Spence, M. A., Westlake, J., and Lange, K. (1977). Estimation of the variance components for dermal ridge count. *Ann. Hum. Genet.* **41**(1), 111–115.

Steinmetz, H., Herzog, A., Schlaug G Huang, Y., and Jäncke, L. (1995). Brain (A) symmetry in monozygotic twins. *Cereb. Cortex* **5**(4), 296–300.

Suddath, R. L., Christison, G. W., Torrey, E. F., Casanova, M. F., and Weinberger, D. R. (1990). Anatomical abnormalities in the brains of monozygotic twins discordant for schizophrenia. *N. Engl. J. Med.* **322**(12), 789–794.

Sullivan, E. V., Pfefferbaum, A., Swan, G. E., and Carmelli, D. (2001). Heritability of hippocampal size in elderly twin men: Equivalent influence from genes and environment. *Hippocampus* **11**(6), 754–762.

Sullivan, P. F., Kendler, K. S., and Neale, M. C. (2003). Schizophrenia as a complex trait: Evidence from a meta-analysis of twin studies. *Arch. Gen. Psychiatry* **60**(12), 1187–1192.

Szeszko, P. R., Lipsky, R., Mentschel, C., Robinson, D., Gunduz-Bruce, H., Sevy, S., Ashtari, M., Napolitano, B., Bilder, R. M., Kane, J. M., Goldman, D., and Malhotra, A. K. (2005). Brain-derived neurotrophic factor val66met polymorphism and volume of the hippocampal formation. *Mol. Psychiatry* **10**(7), 631–636.

Tohgi, H., Takahashi, S., Kato, E., Homma, A., Niina, R., Sasaki, K., Yonezawa, H., and Sasaki, M. (1997). Reduced size of right hippocampus in 39- to 80-year-old normal subjects carrying the apolipoprotein E epsilon4 allele. *Neurosci. Lett.* **236**(1), 21–24.

Thompson, P. M., and Toga, A. W. (1997). Detection, visualization and animation of abnormal anatomic structure with a deformable probabilistic brain atlas based on random vector field transformations. *Med. Image Anal.* **1**(4), 271–294.

Thompson, P. M., Schwartz, C., and Toga, A. W. (1996a). High-resolution random mesh algorithms for creating a probabilistic 3D surface atlas of the human brain. *Neuroimage* **3**(1), 19–34.

Thompson, P. M., Schwartz, C., Lin, R. T., Khan, A. A., and Toga, A. W. (1996b). Three-dimensional statistical analysis of sulcal variability in the human brain. *J. Neurosci.* **16**(13), 4261–4274.

Thompson, P. M., Cannon, T. D., Narr, K., van Erp, T., Poutanen, V. P., Huttunen, M., Lönnqvist, J., Standertskjöld-Nordenstam, C. G., Kaprio, J., Khaledy, M., Dail, R., Zoumalan, C. I., *et al.* (2001). Genetic influences on brain structure. *Nat. Neurosci.* **4**(12), 1253–1258.

Thompson, P., Cannon, T. D., and Toga, A. W. (2002). Mapping genetic influences on human brain structure. *Ann. Med.* **34**(7–8), 523–536. Review.

Tramo, M. J., Loftus, W. C., Stukel, T. A., Green, R. L., Weaver, J. B., and Gazzaniga, M. S. (1998). Brain size, head size, and intelligence quotient in monozygotic twins. *Neurology* **50**(5), 1246–1252.

Tsujita, T., Niikawa, N., Yamashita, H., Imamura, A., Hamada, A., Nakane, Y., and Okazaki, Y. (1998). Genomic discordance between monozygotic twins discordant for schizophrenia. *Am. J. Psychiatry* **155**(3), 422–424.

Uno, H., Tarara, R., Else, J. G., Suleman, M. A., and Sapolsky, R. M. (1989). Hippocampal damage associated with prolonged and fatal stress in primates. *J. Neurosci.* **9**(5), 1705–1711.

van Erp, T. G., Saleh, P. A., Huttunen, M., Lönnqvist, J., Kaprio, J., Salonen, O., Valanne, L., Poutanen, V. P., Standertskjöld-Nordenstam, C. G., and Cannon, T. D. (2004). Hippocampal volumes in schizophrenic twins. *Arch. Gen. Psychiatry* **61**(4), 346–353.

van Haren, N. E., Picchioni, M. M., McDonald, C., Marshall, N., Davis, N., Ribchester, T., Hulshoff Pol, H. E., Sharma, T., Sham, P., Kahn, R. S., and Murray, R. (2004). A controlled study of brain structure in monozygotic twins concordant and discordant for schizophrenia. *Biol. Psychiatry* **56**(6), 454–461.

van Haren, N. E., Hulshoff Pol, H. E., Schnack, H. G., Cahn, W., Mandl, R. C., Collins, D. L., Evans, A. C., and Kahn, R. S. (2007). Focal gray matter changes in schizophrenia across the course of the illness: A 5-year follow-up study. *Neuropsychopharmacology* **32**(10), 2057–2066.

van Haren, N. E., Bakker, S. C., and Kahn, R. S. (2008). Genes and structural brain imaging in schizophrenia. *Curr. Opin. Psychiatry* **21**(2), 161–167. Review.

van Os, J. (2009). A salience dysregulation syndrome. *Br. J. Psychiatry* **194**(2), 101–103.

van Os, J., and Murray, R. (2008). Gene-environment interactions in schizophrenia. Introduction. *Schizophr. Bull.* **34**(6), 1064–1065.

van Os, J., Rutten, B. P., and Poulton, R. (2008). Gene-environment interactions in schizophrenia: Review of epidemiological findings and future directions. *Schizophr. Bull.* **34**(6), 1066–1082.

Verdoux, H., Geddes, J. R., Takei, N., Lawrie, S. M., Bovet, P., Eagles, J. M., Heun, R., McCreadie, R. G., McNeil, T. F., O'Callaghan, E., Stöber, G., Willinger, M. U., *et al.* (1997). Obstetric complications and age at onset in schizophrenia: An international collaborative meta-analysis of individual patient data. *Am. J. Psychiatry* **154**(9), 1220–1227.

Vincent, J. B., Kalsi, G., Klempan, T., Tatuch, Y., Sherrington, R. P., Breschel, T., McInnis, M. G., Brynjolfsson, J., Petursson, H., Gurling, H. M., Gottesman, I. I., Torrey, E. F., *et al.* (1998). No evidence of expansion of CAG or GAA repeats in schizophrenia families and monozygotic twins. *Hum. Genet.* **103**(1), 41–47.

Wallace, G. L., Eric Schmitt, J., Lenroot, R., Viding, E., Ordaz, S., Rosenthal, M. A., Molloy, E. A., Clasen, L. S., Kendler, K. S., Neale, M. C., and Giedd, J. N. (2006). A pediatric twin study of brain morphometry. *J. Child Psychol. Psychiatry* **47**(10), 987–993.

Wang, J. K., Li, Y., and Su, B. (2008). A common SNP of MCPH1 is associated with cranial volume variation in Chinese population. *Hum. Mol. Genet.* **17**(9), 1329–1335.

Ward, K. E., Friedman, L., Wise, A., and Schulz, S. C. (1996). Meta-analysis of brain and cranial size in schizophrenia. *Schizophr. Res.* **22**(3), 197–213.

Wassink, T. H., Nelson, J. J., Crowe, R. R., and Andreasen, N. C. (1999). Heritability of BDNF alleles and their effect on brain morphology in schizophrenia. *Am. J. Med. Genet.* **88**(6), 724–728.

Wassink, T. H., Crowe, R. R., and Andreasen, N. C. (2000). Tumor necrosis factor receptor-II: Heritability and effect on brain morphology in schizophrenia. *Mol. Psychiatry* **5**(6), 678–682.

Wassink, T. H., Nopoulos, P., Pietila, J., Crowe, R. R., and Andreasen, N. C. (2003). NOTCH4 and the frontal lobe in schizophrenia. *Am. J. Med. Genet. B Neuropsychiatr. Genet.* **118B**(1), 1–7.

Weinberger, D. R. (1995). From neuropathology to neurodevelopment. *Lancet* **346**(8974), 552–557. Review.

White, T., Andreasen, N. C., and Nopoulos, P. (2002). Brain volumes and surface morphology in monozygotic twins. *Cereb Cortex* **12**(5), 486–493.

Williams, R. R., Hunt, S. C., Hasstedt, S. J., Hopkins, P. N., Wu, L. L., Berry, T. D., Stults, B. M., Barlow, G. K., and Kuida, H. (1990). Genetics of hypertension: What we know and don't know. *Clin. Exp. Hypertens. A.* **12**(5), 865–876.

Williams, H. J., Owen, M. J., and O'Donovan, M. C. (2007). Is COMT a susceptibility gene for schizophrenia? *Schizophr. Bull.* **33**(3), 635–641.

Woods, R. P., Freimer, N. B., De Young, J. A., Fears, S. C., Sicotte, N. L., Service, S. K., Valentino, D. J., Toga, A. W., and Mazziotta, J. C. (2006). Normal variants of Microcephalin and ASPM do not account for brain size variability. *Hum. Mol. Genet.* **15**(12), 2025–2029.

Wright, I. C., Rabe-Hesketh, S., and Woodruff, P. W. (2000). Meta-analysis of regional brain volumes in schizophrenia. *Am. J. Psychiatry* **157**(1), 16–25.

Wright, I. C., Sham, P., Murray, R. M., Weinberger, D. R., and Bullmore, E. T. (2002). Genetic contributions to regional variability in human brain structure: Methods and preliminary results. *Neuroimage* **17**(1), 256–271.

Zinkstok, J., Schmitz, N., van Amelsvoort, T., de Haan, L., Yusuf, M. A., Baas, F., and Linszen, D. (2007). Genetic variation in COMT and PRODH is associated with brain anatomy in patients with schizophrenia. *Genes Brain Behav.* **7**(1), 61–69.

THE ROLE OF STRIATAL NMDA RECEPTORS IN DRUG ADDICTION

Yao-Ying Ma,* Carlos Cepeda,† and Cai-Lian Cui‡

*Stefan & Shirley Hatos Center for Neuropharmacology, University of California Los Angeles, Los Angeles, California, USA

†Mental Retardation Research Center, Semel Institute for Neuroscience and Human Behavior, University of California Los Angeles, Los Angeles, California, USA

‡Neuroscience Research Institute and Department of Neurobiology, Peking University Health Science Center, People's Republic of China

The past decade has witnessed an impressive accumulation of evidence indicating that the excitatory amino acid glutamate and its receptors, in particular the *N*-methyl-D-aspartate (NMDA) receptor subtype, play an important role in drug addiction. Various lines of research using animal models of drug addiction have demonstrated that drug-induced craving is accompanied by significant upregulation of NR2B subunit expression. Furthermore, selective blockade of NR2B-containing NMDA receptors in the striatum, especially in the nucleus accumbens (NAc) can inhibit drug craving and reinstatement. The purpose of this review is to examine the role of striatal NMDA receptors in drug addiction. After a brief description of glutamatergic innervation and NMDA receptor subunit distribution in the striatum, we discuss potential mechanisms to explain the role of striatal NMDA receptors in drug addiction by elucidating signaling cascades involved in the regulation of subunit expression and redistribution, phosphorylation of receptor subunits, as well as activation of intracellular signals

INTERNATIONAL REVIEW OF NEUROBIOLOGY, VOL. 89
DOI: 10.1016/S0074-7742(09)89006-5

triggered by drug experience. Understanding the mechanisms regulating striatal NMDA receptor changes in drug addiction will provide more specific and rational targets to counteract the deleterious effects of drug addiction.

I. Introduction

Drug addiction is defined primarily on the basis of behavioral abnormalities, that is, loss of control over drug use and compulsive seeking and taking despite the knowledge of its adverse consequences (Nestler, 2004a). The switch from controlled to compulsive drug-seeking and drug-taking behaviors involves many neuroadaptations that occur following repeated drug administration, including molecular adaptations, cell morphology alterations, and synaptic transduction modifications. Although dopamine (DA) and opiate receptor systems have received much attention since the beginning of drug addiction research, recent work has focused more on the role of other receptors such as glutamate and γ-aminobutyric acid (GABA), as well as their interactions with DA. In addition, although clinical work has used DA receptor agonists and antagonists to treat drug-seeking behavior, their side effects preclude widespread use. The finding that other receptor systems are involved in drug addiction opens new therapeutic avenues. In particular, NMDA receptors are becoming increasingly recognized as key players in drug seeking and drug taking. The widespread distribution of NMDA receptors in the brain and their critical roles in a wide range of physiological and pathological processes, including neuronal development, synaptic plasticity, epilepsy, and pain, have attracted the attention of researchers in the field of drug addiction. The neuronal adaptations of either glutamate release or NMDA receptor sensitivity, as well as their behavioral consequences, seem to be related to the administration model and the dose of addicting drugs (Kerdsan *et al.*, 2009; Lu *et al.*, 1999; Simoes *et al.*, 2007). These observations suggest that NMDA receptors may play a critical role in the effects of drug addiction. In this review we are going to concentrate on studies where the primary focus has been on striatal NMDA receptors in drug addiction.

The major hypothesis that guides the current neurobiological research on drug addiction is that chronic drug exposure causes long-lasting neuroadaptive changes in the brain at the molecular and cellular levels, which may contribute to compulsive drug use and relapse (Chao and Nestler, 2004; Nestler, 2004b). It is reasonable to believe that at least some of these molecular/neurochemical alterations are associated with drug-seeking and drug-taking behaviors (Shaham *et al.*, 2003). Since most central NMDA receptors are constituted by NR1/NR2 (A or B) assemblies, and opioids and psychostimulants are the most popular drugs of abuse, this review will focus on the involvement of NR1, NR2A, and NR2B receptor subunits in the addiction to opioids and psychostimulants.

II. NMDA Receptors in the Striatum

A. NMDA Receptors

Glutamate is the main excitatory neurotransmitter in the brain and its effects are mediated through ionotropic and metabotropic receptors. Ionotropic receptor subtypes are named after their preferred agonists: α-amino-3-hydroxy-5-methyl-4-isoxazolepropionic acid receptor (AMPA), kainate, and NMDA (Mori and Mishina, 1995). Most NMDA receptors in the brain are thought to be heteromeric complexes with NR1 as the primary constitutive subunit and NR2 (A-D) as functional subunits that affect channel kinetics and sensitivity (Mori and Mishina, 1995). The NMDA receptor is a ligand-gated channel complex. It is also gated by its membrane depolarization that removes Mg^{2+} block and confers its voltage dependence (Darlison, 1992; Michaelis, 1998; Schoepfer *et al.*, 1994). This dual requirement for channel opening necessitates the integration of both chemical- and voltage-sensing information at the receptor level. Compared to AMPA and kainate receptors, the NMDA receptor channel is highly permeable to Ca^{2+} and is modulated by multiple mechanisms (Marino and Conn, 2002; Mori and Mishina, 1995). It has been shown that NMDA receptors are critically involved not only in many important physiological functions such as synapse formation and synaptic plasticity, but also in neuropathological processes underlying epilepsy (Tzschentke, 2002), ischemia (Hammerman and Kaplan, 1998), pain (De Kock and Lavand'homme, 2007), Parkinson's disease (Bonuccelli and Del Dotto, 2006), Huntington's disease (Fan *et al.*, 2007), and drug addiction (Aguilar *et al.*, 2009; Comer and Sullivan, 2007; Ma *et al.*, 2006, 2007).

B. Glutamatergic Innervation of the Striatum

The striatum is the major input structure of the basal ganglia. The principal neuronal population (comprising more than 90% of all striatal neurons) is constituted by medium-sized spiny neurons (MSSNs), which are GABAergic projection neurons that segregate into two distinct populations (direct and indirect pathways) based primarily on the output areas they innervate and DA receptor expression (Afifi, 1994; Gerfen, 2000; Kreitzer and Malenka, 2008; Smith and Villalba, 2008). The striatum also is topographically organized into patch and matrix compartments based on specific inputs, outputs, and enzyme or receptor content (Gerfen, 1984; Voorn *et al.*, 2004).

MSSNs receive glutamatergic inputs that activate postsynaptic NMDA and non-NMDA receptors. Although both the dorsal and ventral (nucleus accumbens, NAc) parts of striatum receive extensive excitatory afferents from the cerebral cortex and thalamus, a regional difference of glutamatergic innervation originating from the cortex and thalamus has been demonstrated, in particular a dorsolateral-to-ventromedial gradient (Voorn *et al.*, 2004). While most of the dorsolateral striatum is innervated primarily by sensory and motor cortical areas, the most ventromedial part predominantly receives visceral-associated information, and the striatal region between these two areas receives information from higher associational cortical areas (Voorn *et al.*, 2004). Typically, the dorsal striatum receives glutamatergic projections from practically all cerebral cortical areas, including the sensorimotor cortex, dorsal anterior cingulate cortex, and dorsal prelimbic cortex; whereas the ventral striatum receives inputs from deeper layers of the medial prefrontal cortex (mPFC). In terms of its two major territories, the core component of the NAc receives cortical input predominantly from both dorsal and ventral parts of prelimbic, whereas the shell component of the NAc is innervated primarily by ventral parts of the prelimbic cortex and also the inferolimbic cortex (reviewed by Sesack and Grace, 2009).

The functions and origin of thalamostriatal projections are less known as compared to the corticostriatal pathways. From lateral to medial, the dorsal striatum receives glutamatergic innervations from the central, paracentral, and central medial thalamic nucleus. In the ventral striatum, the shell component of the NAc receives excitatory projections primarily from the paraventricular thalamic nucleus, whereas the core component of the NAc is innervated primarily by the intermediodorsal thalamic nucleus (Voorn *et al.*, 2004).

The amygdalostriatal projections, as well as the hippocampostriatal innervations, are specifically observed in the ventral part of the striatum and "taper off" in a dorsolateral direction (Voorn *et al.*, 2004). The amygdala, especially the basolateral component, supplies the NAc shell heavily, but only reaches the patch compartments in the core region of the NAc. Furthermore, the shell compartment receives dense ventral hippocampus projections, whereas the core component is innervated mostly by the dorsal parts of the subiculum (Meredith *et al.*, 2008)

C. Distribution of NMDA Receptors in Striatum

The NR1 subunit family, including eight splice variants of a single gene, is distributed ubiquitously in the brain. The NR2 subunit family, encoded by four distinct genes (NR2A, NR2B, NR2C, NR2D), shows a characteristic distribution in the brain, with patterns of expression that change during development

(Monyer *et al.*, 1994; Standaert *et al.*, 1996; Watanabe *et al.*, 1994). *In situ* hybridization analyses showed a high expression of NR2A and NR2B, but low, if any, mRNA levels for NR2C and NR2D in striatum (Kutsuwada *et al.*, 1992; Monyer *et al.*, 1994; Mori and Mishina, 1995; Watanabe *et al.*, 1993).

In human and rat striatum, projection neurons express NR1, NR2A, and NR2B subunits; whereas interneurons contain NR1, NR2B, and low levels of NR2D mRNAs (Kuppenbender *et al.*, 2000). The most striking difference in neuronal subunit distribution between humans and rats is the lack of NR2A mRNA in human striatal somatostatin (SOM)-positive interneurons, but abundant NR2A mRNA signal is detected in SOM-positive neurons of rats (Kuppenbender *et al.*, 2000). Using biochemical protein fractionation and coimmunoprecipitation approaches, Dunah and Standaert (2003) determined the subcellular segregation of distinct heteromeric NMDA receptors in the striatum. NMDA receptors in the endoplasmic reticulum and synaptic vesicle membrane are primarily binary complexes (NR1/NR2A and NR1/NR2B). In contrast, a majority of NMDA receptors in the postsynaptic membrane are ternary, containing all three subunits (NR1/NR2A/NR2B). This suggests differential targeting of NMDA receptor assemblies between intracellular and postsynaptic sites based on subunit composition (Dunah and Standaert, 2003). It also has been postulated that the binary assemblies represent partially assembled receptor intermediates, or that they are targeted for degradation.

There is also a gradient of NMDA receptor subunit expression and function in striatum. Thus, although NR2B subunit expression is uniform throughout the striatum, NR2A subunit expression is greater in the lateral areas of the striatum. Electrophysiological recordings demonstrated that the decay time of NMDA receptor mediated excitatory postsynaptic currents is significantly slower in MSSNs from ventromedial as compared to the dorsolateral striatum, suggesting that the synaptic processing of glutamatergic inputs is different in the ventromedial and dorsolateral striatum of adult rat brain (Chapman *et al.*, 2003).

III. Role of Glutamate and NMDA Receptors in Drug Addiction

A. NMDA Receptor Subunit Expression in Drug Addiction

The protein level of NMDA receptor subunits is one of most effective measures of glutamatergic synaptic strength for striatal projection neurons (Cowan *et al.*, 2008; Estrada-Sanchez *et al.*, 2009). The positive association between expression of NMDA receptor subunits, especially in the membrane compartment of the synaptosome, with addictive behaviors has been demonstrated by researches focusing on opioids as well as psychostimulants.

Using the conditioned place preference (CPP) model, in which animals are reinforced by drug rewarding effects in a specific environment thereby showing preference to this environment, a robust increase of NR2B, but not NR2A and NR1, subunits was detected in the NAc from rats conditioned by morphine. However, no changes in NMDA receptor subunit expression were observed in rats conditioned by natural rewards such as food, novelty, and social interaction (Ma *et al.*, 2006). Relapse to opioid administration following prolonged periods of abstinence remains the most difficult challenge for addiction treatment. Our own studies demonstrated an overexpression of NR2B subunits in the NAc of rats with a reinstated place preference to the previous morphine-copaired context (Ma *et al.*, 2007). In this experiment, the extinguished place preference was reinstated by either a priming injection of morphine or an acute stress produced by 10 min of forced swimming (FS). An overexpression of NR2B subunits in the hippocampus, however, was only observed in reinstatement induced by morphine, but not by FS. It was thus proposed that an upregulation of NR2B subunits in the NAc is a common phenomenon in the reinstatement of CPP, while in the hippocampus it may be relatively specific for morphine-induced reinstatement.

NMDA receptor subunit expression after psychostimulant treatment seems dependent on the model of drug administration. In animals trained and motivated to self-administration of drugs, NR1 gene expression is significantly increased as compared to noncontingent cocaine exposure. Animals with a previous contingent cocaine use history, showed a gradual recovery of NR1 gene expression in forebrain regions, including dorsal and ventral striatum (Crespo *et al.*, 2002).

It has been shown that noncontingent cocaine exposure, in which animals receive drug injections from the experimenter and are assessed by behavioral sensitization, increases NMDA receptor expression in the striatum. With a treatment of repeated cocaine injections by the experimenter, upregulation of NR1, NR2B subunit expression in striatum is more frequently reported (Scheggi *et al.*, 2002; Schumann and Yaka, 2009). Schumann and Yaka (2009) also found an upregulation of NR2A subunits in cocaine-sensitized rodents. However, it was shown that at 14 days of withdrawal from cocaine, there was a significant (~27%) decrease in the expression of the NR1 C2-containing splice variants in the striatum (Loftis and Janowsky, 2002). The downregulation of this specific NR1 splice variant was also detected in frontal cortex and hippocampus. It is suggested that the study of NR1 subunits in addiction research needs to be extended to specific splice variants.

Furthermore, effects of psychostimulants on NMDA receptor subunit expression also depend on the withdrawal period from drugs. Schumann and Yaka found that prolonged (21 days) but not short (1 day) withdrawal from repeated cocaine injections increases the protein levels of NR1, NR2A, and NR2B in the synaptosomal membranes as well as in the total tissue homogenates from the NAc.

This indicates an increased protein synthesis of the NMDA receptor subunits (Schumann and Yaka, 2009). Similarly, at about the same time, Ghasemzadeh *et al.* found that repeated cocaine injections are associated with upregulation of NR2B subunits in dorsolateral striatum only after an extended withdrawal period (Ghasemzadeh *et al.*, 2009). Also, the decreased expression of the NR1 C2-containing splice variants in the striatum was only observed at 14 days, but not 1 day after cocaine injections (Loftis and Janowsky, 2002).

Other studies, however, showed that noncontingent injections of psychostimulants, either cocaine or amphetamine, were disassociated with any change in NMDA receptor subunit expression. Experiments by Sun *et al.* (2008), using striatal neurons cocultured with prefrontal cortex neurons, demonstrated that either acute or chronic cocaine treatment did not change striatal surface expression of NMDA receptor subunits. Also Nelson *et al.* (2009) found no significant changes in NR2A and NR2B subunit expression following repeated amphetamine administration.

B. NMDA Receptor Trafficking and Localization

Using the protein crosslinking reagent bis-(sulfuccinimidyl) suberate combined with Western blotting, Nelson *et al.* (2009) found no significant changes in the ratio of NR2A/NR2B protein level in the surface as compared to that in the intracellular pools in both the dorsolateral striatum and the NAc following repeated amphetamine administration. It is suggested that there is no redistribution of NR2 subunits induced by chronic treatment of amphetamine.

It appears that NMDA receptor trafficking and redistribution also depend on the withdrawal period from drugs. Although no upregulated protein synthesis of the NMDA receptor subunits occurs 1 day after withdrawal from repeated cocaine injections, Schumann and Yaka observed an increase of NR1, NR2A, and NR2B only in the light membrane fraction (representing the intracellular membrane content), but not in the total tissue homogenates of the NAc shortly after cocaine withdrawal. These results suggest that NMDA receptor subunits were redistributed within the cells and not degraded in the NAc 1 day after cocaine treatment (Schumann and Yaka, 2009).

As opposed to substantial evidence regarding psychostimulant-induced NMDA receptor trafficking and redistribution, much less is known about the effects of opioid treatment on striatal neurons. Although our previous experiments found that membrane NR2B protein levels, but not NR1 or NR2A, were robustly increased during the expression and reinstatement of morphine-conditioned CPP, no direct tests to find the subunit ratio of protein levels between the intracellular and membrane compartments were performed (Ma *et al.*, 2006, 2007). More studies are needed to determine if NMDA receptor subunit trafficking is regulated by opioid treatment.

C. NMDA Receptor Phosphorylation

NMDA receptors can be modulated by protein kinases, such as protein kinase A (PKA), protein kinase C (PKC), and Ca^{2+}/calmodulin-dependent protein kinases (CaMKs), as well as protein phosphatases (Dingledine *et al.*, 1999; Wang *et al.*, 2006). It was reported that threonine 879 and serines 890 and 896 at the C-terminus of the NR1 subunit can be phosphorylated by PKC, whereas serine 897 is phosphorylated by PKA. The NR2B C-terminus is phosphorylated at serine 1303 by CaMKII and PKC, and serine 1323 by PKC. Furthermore, NR2 subunits, especially NR2B in the striatum, can be phosphorylated at tyrosine sites by Src-family tryrosine kinases (Nakazawa *et al.*, 2001; Suzuki and Okumura-Noji, 1995; Zheng *et al.*, 1998). The phosphorylation of NMDA receptors at either tyrosine or serine sites may facilitate surface membrane insertion, synaptic delivery and endocytosis (Dunah *et al.*, 2004; Hall and Soderling, 1997), increased NMDA receptor dependent excitatory postsynaptic currents, and thus increased synaptic strength (see review by Wang *et al.*, 2006).

Phosphor- and site-specific antibodies against the specific phosphorylation site on the NR1 or NR2 subunit was used to test phosphorylation modifications in NMDA receptor subunits after administration of drugs of abuse. *In vitro* studies by Sanchez-Blazquez *et al.* demonstrated that morphine treatment provoked the phosphorylation of specific cytosolic tyrosine residues in the NR2A subunit by activating PKC gamma isoform and therefore activating Src, enabling Src to phosphorylate the Tyr1325 residue of NR2A subunits (Sanchez-Blazquez *et al.*, 2009). Activation of D1 receptors induces the phosphorylation of Ser897 at the NR1 subunit by PKA and therefore enhances the phosphorylation of cAMP response element binding protein (CREB). DA is unable to induce cAMP response elements (CRE) mediated target gene expression in striatal neurons transfected with a phosphorylation-deficient NR1 construct (Dudman *et al.*, 2003). *In vitro* data thus suggest that both opioids and psychostimulants can modulate the phosphorylation of NMDA receptor subunits.

The modulation of NMDA receptor phosphorylation by drugs was also demonstrated by *in vivo* studies. Using the CPP model for evaluating rewarding effects of drugs, Kato *et al.* (2007) found that morphine conditioning increased the phosphorylating level of Tyr 1472 of NR2B subunits in the mouse limbic forebrain containing the NAc, and that activation of Src family kinase in the limbic forebrain may be involved in this phosphorylation. Similar to opioids, psychostimulants phosphorylate NMDA receptor subunits in the striatum (Kato *et al.*, 2007). Repeated cocaine injection (20 mg/kg) for 9 days increases the phosphorylation at Ser897 of NR1 subunits in the dorsal striatum, which is blocked by injection of antagonists to either DA D1 or NMDA receptors. It is suggested that D1 receptor stimulation in response to the chronic cocaine treatments is responsible for phosphorylating NR1 subunits in the dorsal striatum (Lee *et al.*, 2008). Acute amphetamine administration significantly enhances NR1 serine 896 phosphorylation in D1 receptor expressing dynorphin-containing striatonigral projection neurons forming the direct pathway (Liu *et al.*, 2004). The *in vitro* data reported by

Dudman *et al.* (2003) suggest that NR1, which is an obligatory NMDA receptor subunit, also is involved in modulating NMDA receptor function.

D. Interaction Between DA and NMDA Receptors

DA and NMDA receptors interact in striatal MSSNs. The outcome of these interactions depends on a number of variables, in particular the DA receptor subtype (Cepeda and Levine, 1998). In both dorsal and ventral striatum, D1 receptor activation enhances NMDA receptor mediated responses via various signaling cascades (Cepeda *et al.*, 1998; Chergui and Lacey, 1999; Harvey and Lacey, 1997). In contrast, D2 receptor activation decreases NMDA responses (Cepeda *et al.*, 1993). This suggests a possible reinforcing mechanism whereby D1 receptor-containing MSSNs in dorsal and ventral striatum, when primed by phasic DA release induced by rewards, produce an increment of NMDA responses making coincident glutamatergic inputs more relevant. This coincidence detection results from the integration of both the DA and glutamate signaling pathways. These DA and NMDA receptor interactions could contribute significantly to synaptic strengthening and the reinforcing effects of drugs.

Modulation of NMDA receptor phosphorylation has been accepted as one important mechanism in DA and NMDA receptor interactions after acute or chronic treatment by drugs of abuse. Activation of D1 receptors expressed in striatal neurons facilitates the membrane insertion of NR2A/NR2B from the intracellular compartment into the postsynaptic membrane (Dunah *et al.*, 2004) and the clustering of NR2B subunits along dendritic spine shafts of striatal MSSNs (Hallett *et al.*, 2006). These effects can be blocked by the protein tyrosine kinase inhibitor genistein (Dunah *et al.*, 2004), and also can be mimicked by the protein phophatase inhibitor pervanadate (Hallett *et al.*, 2006). This strongly suggests that phosphorylation of NMDA receptor subunits is critical for the interaction between D1 receptors and NMDA receptors. However, activation of D2 receptors disrupts the association of CaMKs with NR2B and thus decreases NR2B phosphorylation (Liu *et al.*, 2006).

Some intracellular events, including activation of DA and cyclic AMP-regulated phosphoprotein-32 (DARPP-32) and extracellular signal-regulated kinase (ERK), have been shown to depend on the stimulation of both DA, especially D1, and NMDA receptors, in particular those expressing NR2B subunits (Schumann and Yaka, 2009; Svenningsson *et al.*, 2004; Valjent *et al.*, 2005).

E. Intracellular Signaling

To unveil how drugs of abuse induce permanent, compulsive behavior, it is necessary to identify the intracellular events that trigger gene transcription alterations supporting long-lasting synaptic changes (Hyman *et al.*, 2006; Reynolds and Wickens, 2002; Stipanovich *et al.*, 2008). Intracellular adaptations in response to administration of drugs are involved, at least partially, in the regulation and transduction of NMDA receptor signals (Scheggi *et al.*, 2004).

DARPP-32, which is highly enriched in striatal MSSNs, is a prominent mediator of rewarding signaling. DARPP-32 has been detected in both cytoplasmic and nuclear components of striatal GABAergic neurons (Stipanovich *et al.*, 2008). The function of DARPP-32 is determined by its phosphorylation state. D1 receptor-mediated activation of PKA can phosphorylate DARPP-32 at threonine 34 and thereafter convert DARPP-32 into a protein inhibitor of the multifunctional serine/threonine protein phosphatase-1 (PP1). Svenningsson *et al.* (2004) found that DARPP-32 mediated inhibition of PP1 increases the phosphorylation of neurotransmitter receptors, including NMDA receptors and ion channels. Cyclin-dependent kinase 5 (Cdk5)-mediated phosphorylation of DARPP-32 at threonine 75 converts DARPP-32 into an inhibitor of PLA and thereby inhibits phosphorylation of NMDA receptors as well as AMPA receptors (Bibb *et al.*, 1999). Scheggi *et al.* (2004) found that the phosphorylation levels of NR1 subunit decreased in parallel with those of DARPP-32 at threonine-75 site, supporting the hypothesis that the phosphorylating status of DARPP-32 is responsible for phosphorylation of NMDA receptor subunits and thereby regulates the location and activation of NMDA receptors.

The potential role of ERK in drug addiction has attracted much attention in the past decade. Pharmacological blockade of ERK attenuates or abolishes the rewarding effects of cocaine and morphine (Ozaki *et al.*, 2004; Valjent *et al.*, 2000, 2006). ERK activity is also known as one important contributor to long-term synaptic plasticity (Thomas and Huganir, 2004; Valjent *et al.*, 2005). ERK may work as an intracellular codetector of both DA and glutamate receptor activation. Valjent *et al.* found that activation of ERK requires the stimulation of both D1 and NMDA receptors in striatal neurons. Further studies showed that activation of ERK through combined action of NMDA and D1 receptors was absent in mice lacking DARPP-32 (Valjent *et al.*, 2005). Inhibition of NR2B-containing NMDA receptors decreased ERK activity in the NAc and also prevented psychomotor sensitization induced by repeated cocaine injections (Schumann and Yaka, 2009). Given the phosphorylation of NMDA receptors as well as activation of intracellular kinases targeting the threonine/serine or tyrosine sites, a phosphatase cascade is strongly suggested to participate in mediating NMDA receptor role in drug addiction.

F. Generation of Silent Synapses by Drug Experience

Silent synapses, enriched in the developing brain, are excitatory glutamatergic synapses and exhibit NMDA-receptor-mediated responses, but no AMPA receptor-mediated responses. Studies over the last decade have demonstrated that silencing of nonsilent synapses is a prominent cellular mechanism of synaptic plasticity in early postnatal development, but synapse elimination can also occur in the mature brain (Huang *et al.*, 2009). The initial number of synapses formed in the brain is much larger than the number of synapses retained, which suggests that synapse elimination is a crucial step and consequence of experience-dependent learning.

Very recently, Huang *et al.* (2009) reported that *in vivo* cocaine experience *de novo* induced silent synapses, which was mediated specifically by membrane insertions of new, NR2B-containing NMDA receptors. Our previous data showed an upregulation of NR2B subunit expression in the NAc and an inhibitory effect of microinjection of an NR2B-selective NMDA receptor antagonist into the NAc shell component on the development, expression, and reinstatement of CPP induced by morphine. If the new membrane insertion of NR2B-containing NMDA receptors in MSSNs is the exact substrate for silent synapses induced by drugs of abuse, we proposed that inhibiting NR2B-containing NMDA receptors disabled synapse elimination induced by *in vivo* morphine experience or morphine-associated context. These generated silent synapses may further contribute to the presence of NAc MSSNs with an increased efficiency for recruiting AMPARs to strengthen excitatory synaptic transmission.

IV. Neuropharmacological Studies and Clinical Implications

A large body of research has shown that pharmacological blockade of NMDA receptors inhibits acquisition, expression, as well as reinstatement of drug craving. In contrast to unreliable effects on drug craving for psychostimulants, positive effects of NMDA receptor antagonists have been demonstrated for opioid addiction.

Studies in our laboratory demonstrated that ifenprodil, a NMDA receptor antagonist highly selective for NR2B subunits, produced a dose-dependent reduction in reward craving induced by morphine in a dose-dependent manner, with little or no effect on the reward craving induced by food consumption and social interaction (Ma *et al.*, 2006). It is interesting that ifenprodil also prevented the development and expression of place preference induced by a novel environment. It has been suggested that there is a very close interrelation between the behavioral and physiological mechanisms underlying an animal's sensitivity to novelty as well as to drugs of abuse (Bevins *et al.*, 2002; Cervo and Samanin, 1996; Kim *et al.*, 2004). An animal's sensitivity to novelty appears to be a powerful predictive variable for later sensitivity to drugs of abuse. The blockade of craving for novelty is speculated to decrease the craving for drugs of abuse. Furthermore, we observed a dose-dependent reduction of morphine- but not FS-induced reinstatement after systemic injection of ifenprodil. It is suggested that the neural mechanisms mediating drug-induced reinstatement are NMDA receptor dependent, which is dissociable from the molecular/neurochemical substrates mediating stress-induced reinstatement. Microinjection of ifenprodil directly into the NAc shell inhibited the reinstating effect of morphine priming. The aim of the pharmacotherapy for drug addiction is to find a way to selectively interfere with the drug-related responses, while leaving other biologically important behaviors such as food consumption, novelty approach, and social interaction intact. One of the commonly accepted views on the mechanisms of compulsive drug seeking and taking is that both drug and natural rewards are mediated by a common neuroanatomical pathway. A potential use of ifenprodil is indicated for the treatment of opioid addicts. First, as ifenprodil itself does not support CPP, the probability of abusing this antagonist is low. Second, ifenprodil does not seem to affect the place preference induced by natural rewards, such as food consumption and social interaction. Third, considering its potent inhibitory effect on the development and expression of morphine-induced CPP, ifenprodil is expected to prevent the relapse to opioid drugs. Finally, psychostimulant- and psychotomimetic-like effects, frequently occurring with nonselective NMDA receptor antagonists, have never been reported when using ifenprodil (Narita *et al.*, 2000). This combination of effects could prove a promising maneuver serving not only as an adjunct to therapy in counteracting further development of addiction, but also as a means of preventing relapse to drugs.

Acknowledgment

This work was supported by a grant (DA05010) from NIH, USA. The work of Ma *et al.* presented in this review was supported by the National Basic Research Program (2003-CB515407) of China and Center of Excellence Grant (P01 AT-002038-01A1) from NIH, USA.

References

Afifi, A. K. (1994). Basal ganglia: Functional anatomy and physiology. Part 2. *J. Child Neurol.* **9,** 352–361.

Aguilar, M. A., Manzanedo, C., Do Couto, B. R., Rodriguez-Arias, M., and Minarro, J. (2009). Memantine blocks sensitization to the rewarding effects of morphine. *Brain Res.* **1288,** 95–104.

Bevins, R. A., Besheer, J., Palmatier, M. I., Jensen, H. C., Pickett, K. S., and Eurek, S. (2002). Novel-object place conditioning: Behavioral and dopaminergic processes in expression of novelty reward. *Behav. Brain Res.* **129,** 41–50.

Bibb, J. A., Snyder, G. L., Nishi, A., Yan, Z., Meijer, L., Fienberg, A. A., Tsai, L. H., Kwon, Y. T., Girault, J. A., Czernik, A. J., Huganir, R. L., Hemmings, H. C. Jr., *et al.* (1999). Phosphorylation of DARPP-32 by Cdk5 modulates dopamine signalling in neurons. *Nature* **402,** 669–671.

Bonuccelli, U., and Del Dotto, P. (2006). New pharmacologic horizons in the treatment of Parkinson disease. *Neurology* **67,** S30–S38.

Cepeda, C., Buchwald, N. A., and Levine, M. S. (1993). Neuromodulatory actions of dopamine in the neostriatum are dependent upon the excitatory amino acid receptor subtypes activated. *Proc. Natl. Acad. Sci. USA* **90,** 9576–9580.

Cepeda, C., and Levine, M. S. (1998). Dopamine and N-methyl-D-aspartate receptor interactions in the neostriatum. *Dev. Neurosci.* **20,** 1–18.

Cepeda, C., Colwell, C. S., Itri, J. N., Chandler, S. H., and Levine, M. S. (1998). Dopaminergic modulation of NMDA-induced whole cell currents in neostriatal neurons in slices: Contribution of calcium conductances. *J. Neurophysiol.* **79,** 82–94.

Cervo, L., and Samanin, R. (1996). Effects of dopaminergic and glutamatergic receptor antagonists on the establishment and expression of conditioned locomotion to cocaine in rats. *Brain Res.* **731,** 31–38.

Chao, J., and Nestler, E. J. (2004). Molecular neurobiology of drug addiction. *Annu. Rev. Med.* **55,** 113–132.

Chapman, D. E., Keefe, K. A., and Wilcox, K. S. (2003). Evidence for functionally distinct synaptic NMDA receptors in ventromedial versus dorsolateral striatum. *J. Neurophysiol.* **89,** 69–80.

Chergui, K., and Lacey, M. G. (1999). Modulation by dopamine D1-like receptors of synaptic transmission and NMDA receptors in rat nucleus accumbens is attenuated by the protein kinase C inhibitor Ro 32-0432. *Neuropharmacology* **38,** 223–231.

Comer, S. D., and Sullivan, M. A. (2007). Memantine produces modest reductions in heroin-induced subjective responses in human research volunteers. *Psychopharmacology (Berl)* **193,** 235–245.

Cowan, C. M., Fan, M. M., Fan, J., Shehadeh, J., Zhang, L. Y., Graham, R. K., Hayden, M. R., and Raymond, L. A. (2008). Polyglutamine-modulated striatal calpain activity in YAC transgenic huntington disease mouse model: Impact on NMDA receptor function and toxicity. *J. Neurosci.* **28,** 12725–12735.

Crespo, J. A., Oliva, J. M., Ghasemzadeh, M. B., Kalivas, P. W., and Ambrosio, E. (2002). Neuroadaptive changes in NMDAR1 gene expression after extinction of cocaine self-administration. *Ann. N. Y. Acad. Sci.* **965,** 78–91.

Darlison, M. G. (1992). Invertebrate GABA and glutamate receptors: Molecular biology reveals predictable structures but some unusual pharmacologies. *Trends Neurosci.* **15,** 469–474.

De Kock, M. F., and Lavand'homme, P. M. (2007). The clinical role of NMDA receptor antagonists for the treatment of postoperative pain. *Best Pract. Res. Clin. Anaesthesiol.* **21,** 85–98.

Dingledine, R., Borges, K., Bowie, D., and Traynelis, S. F. (1999). The glutamate receptor ion channels. *Pharmacol. Rev.* **51,** 7–61.

Dudman, J. T., Eaton, M. E., Rajadhyaksha, A., Macias, W., Taher, M., Barczak, A., Kameyama, K., Huganir, R., and Konradi, C. (2003). Dopamine D1 receptors mediate CREB phosphorylation via phosphorylation of the NMDA receptor at Ser897-NR1. *J. Neurochem.* **87,** 922–934.

Dunah, A. W., and Standaert, D. G. (2003). Subcellular segregation of distinct heteromeric NMDA glutamate receptors in the striatum. *J. Neurochem.* **85,** 935–943.

Dunah, A. W., Sirianni, A. C., Fienberg, A. A., Bastia, E., Schwarzschild, M. A., and Standaert, D. G. (2004). Dopamine D1-dependent trafficking of striatal N-methyl-D-aspartate glutamate receptors requires Fyn protein tyrosine kinase but not DARPP-32. *Mol. Pharmacol.* **65,** 121–129.

Estrada-Sanchez, A. M., Montiel, T., Segovia, J., and Massieu, L. (2009). Glutamate toxicity in the striatum of the R6/2 Huntington's disease transgenic mice is age-dependent and correlates with decreased levels of glutamate transporters. *Neurobiol. Dis.* **34,** 78–86.

Fan, M. M., Fernandes, H. B., Zhang, L. Y., Hayden, M. R., and Raymond, L. A. (2007). Altered NMDA receptor trafficking in a yeast artificial chromosome transgenic mouse model of Huntington's disease. *J. Neurosci.* **27,** 3768–3779.

Gerfen, C. R. (1984). The neostriatal mosaic: Compartmentalization of corticostriatal input and striatonigral output systems. *Nature* **311,** 461–464.

Gerfen, C. R. (2000). Molecular effects of dopamine on striatal-projection pathways. *Trends Neurosci.* **23,** S64–S70.

Ghasemzadeh, M. B., Mueller, C., and Vasudevan, P. (2009). Behavioral sensitization to cocaine is associated with increased glutamate receptor trafficking to the postsynaptic density after extended withdrawal period. *Neuroscience* **159,** 414–426.

Hall, R. A., and Soderling, T. R. (1997). Differential surface expression and phosphorylation of the N-methyl-D-aspartate receptor subunits NR1 and NR2 in cultured hippocampal neurons. *J. Biol. Chem.* **272,** 4135–4140.

Hallett, P. J., Spoelgen, R., Hyman, B. T., Standaert, D. G., and Dunah, A. W. (2006). Dopamine D1 activation potentiates striatal NMDA receptors by tyrosine phosphorylation-dependent subunit trafficking. *J. Neurosci.* **26,** 4690–4700.

Hammerman, C., and Kaplan, M. (1998). Ischemia and reperfusion injury. The ultimate pathophysiologic paradox. *Clin. Perinatol.* **25,** 757–777.

Harvey, J., and Lacey, M. G. (1997). A postsynaptic interaction between dopamine D1 and NMDA receptors promotes presynaptic inhibition in the rat nucleus accumbens via adenosine release. *J. Neurosci.* **17,** 5271–5280.

Huang, Y. H., Lin, Y., Mu, P., Lee, B. R., Brown, T. E., Wayman, G., Marie, H., Liu, W., Yan, Z., Sorg, B. A., Schluter, O. M., Zukin, R. S., *et al.* (2009). *In vivo* cocaine experience generates silent synapses. *Neuron* **63,** 40–47.

Hyman, S. E., Malenka, R. C., and Nestler, E. J. (2006). Neural mechanisms of addiction: The role of reward-related learning and memory. *Annu. Rev. Neurosci.* **29,** 565–598.

Kato, H., Narita, M., Suzuki, M., Yoshimoto, K., Yasuhara, M., and Suzuki, T. (2007). Role of tyrosine kinase-dependent phosphorylation of NR2B subunit-containing NMDA receptor in morphine reward. *Nihon Arukoru Yakubutsu Igakkai Zasshi* **42,** 13–20.

Kerdsan, W., Thanoi, S., and Nudmamud-Thanoi, S. (2009). Changes in glutamate/NMDA receptor subunit 1 expression in rat brain after acute and subacute exposure to methamphetamine. *J. Biomed. Biotechnol.* **2009,** 329631.

Kim, J. A., Pollak, K. A., Hjelmstad, G. O., and Fields, H. L. (2004). A single cocaine exposure enhances both opioid reward and aversion through a ventral tegmental area-dependent mechanism. *Proc. Natl. Acad. Sci. USA* **101,** 5664–5669.

Kreitzer, A. C., and Malenka, R. C. (2008). Striatal plasticity and basal ganglia circuit function. *Neuron* **60,** 543–554.

Kuppenbender, K. D., Standaert, D. G., Feuerstein, T. J., Penney, J. B. Jr., Young, A. B., and Landwehrmeyer, G. B. (2000). Expression of NMDA receptor subunit mRNAs in neurochemically identified projection and interneurons in the human striatum. *J. Comp. Neurol.* **419,** 407–421.

Kutsuwada, T., Kashiwabuchi, N., Mori, H., Sakimura, K., Kushiya, E., Araki, K., Meguro, H., Masaki, H., Kumanishi, T., Arakawa, M., *et al.* (1992). Molecular diversity of the NMDA receptor channel. *Nature* **358,** 36–41.

Lee, D. K., Bian, S., Rahman, M. A., Shim, Y. B., Shim, I., and Choe, E. S. (2008). Repeated cocaine administration increases N-methyl-d-aspartate NR1 subunit, extracellular signal-regulated kinase and cyclic AMP response element-binding protein phosphorylation and glutamate release in the rat dorsal striatum. *Eur. J. Pharmacol.* **590,** 157–162.

Liu, X. Y., Chu, X. P., Mao, L. M., Wang, M., Lan, H. X., Li, M. H., Zhang, G. C., Parelkar, N. K., Fibuch, E. E., Haines, M., Neve, K. A., Liu, F., *et al.* (2006). Modulation of D2R-NR2B interactions in response to cocaine. *Neuron* **52,** 897–909.

Liu, Z., Mao, L., Parelkar, N. K., Tang, Q., Samdani, S., and Wang, J. Q. (2004). Distinct expression of phosphorylated N-methyl-D-aspartate receptor NR1 subunits by projection neurons and interneurons in the striatum of normal and amphetamine-treated rats. *J. Comp. Neurol.* **474,** 393–406.

Loftis, J. M., and Janowsky, A. (2002). Cocaine treatment- and withdrawal-induced alterations in the expression and serine phosphorylation of the NR1 NMDA receptor subunit. *Psychopharmacology (Berl)* **164,** 349–359.

Lu, W., Monteggia, L. M., and Wolf, M. E. (1999). Withdrawal from repeated amphetamine administration reduces NMDAR1 expression in the rat substantia nigra, nucleus accumbens and medial prefrontal cortex. *Eur. J. NeuroSci.* **11,** 3167–3177.

Ma, Y. Y., Chu, N. N., Guo, C. Y., Han, J. S., and Cui, C. L. (2007). NR2B-containing NMDA receptor is required for morphine-but not stress-induced reinstatement. *Exp. Neurol.* **203,** 309–319.

Ma, Y. Y., Guo, C. Y., Yu, P., Lee, D. Y., Han, J. S., and Cui, C. L. (2006). The role of NR2B containing NMDA receptor in place preference conditioned with morphine and natural reinforcers in rats. *Exp. Neurol.* **200,** 343–355.

Marino, M. J., and Conn, P. J. (2002). Direct and indirect modulation of the N-methyl D-aspartate receptor. *Curr. Drug Targets CNS Neurol. Disord.* **1,** 1–16.

Meredith, G. E., Baldo, B. A., Andrezjewski, M. E., and Kelley, A. E. (2008). The structural basis for mapping behavior onto the ventral striatum and its subdivisions. *Brain Struct. Funct.* **213,** 17–27.

Michaelis, E. K. (1998). Molecular biology of glutamate receptors in the central nervous system and their role in excitotoxicity, oxidative stress and aging. *Prog. Neurobiol.* **54,** 369–415.

Monyer, H., Burnashev, N., Laurie, D. J., Sakmann, B., and Seeburg, P. H. (1994). Developmental and regional expression in the rat brain and functional properties of four NMDA receptors. *Neuron* **12,** 529–540.

Mori, H., and Mishina, M. (1995). Structure and function of the NMDA receptor channel. *Neuropharmacology* **34,** 1219–1237.

Nakazawa, T., Komai, S., Tezuka, T., Hisatsune, C., Umemori, H., Semba, K., Mishina, M., Manabe, T., and Yamamoto, T. (2001). Characterization of Fyn-mediated tyrosine phosphorylation sites on GluR epsilon 2 (NR2B) subunit of the N-methyl-D-aspartate receptor. *J. Biol. Chem.* **276,** 693–699.

Narita, M., Aoki, T., and Suzuki, T. (2000). Molecular evidence for the involvement of NR2B subunit containing N-methyl-D-aspartate receptors in the development of morphine-induced place preference. *Neuroscience* **101,** 601–606.

Nelson, C. L., Milovanovic, M., Wetter, J. B., Ford, K. A., and Wolf, M. E. (2009). Behavioral sensitization to amphetamine is not accompanied by changes in glutamate receptor surface expression in the rat nucleus accumbens. *J. Neurochem.* **109,** 35–51.

Nestler, E. J. (2004a). Historical review: Molecular and cellular mechanisms of opiate and cocaine addiction. *Trends Pharmacol. Sci.* **25,** 210–218.

Nestler, E. J. (2004b). Molecular mechanisms of drug addiction. *Neuropharmacology* **47**(Suppl. 1), 24–32.

Ozaki, S., Narita, M., Ozaki, M., Khotib, J., and Suzuki, T. (2004). Role of extracellular signal-regulated kinase in the ventral tegmental area in the suppression of the morphine-induced rewarding effect in mice with sciatic nerve ligation. *J. Neurochem.* **88,** 1389–1397.

Reynolds, J. N., and Wickens, J. R. (2002). Dopamine-dependent plasticity of corticostriatal synapses. *Neural Netw.* **15,** 507–521.

Sanchez-Blazquez, P., Rodriguez-Munoz, M., de la Torre-Madrid, E., and Garzon, J. (2009). Brain-specific Galphaz interacts with Src tyrosine kinase to regulate Mu-opioid receptor-NMDAR signaling pathway. *Cell. Signal.* **21,** 1444–1454.

Scheggi, S., Mangiavacchi, S., Masi, F., Gambarana, C., Tagliamonte, A., and De Montis, M. G. (2002). Dizocilpine infusion has a different effect in the development of morphine and cocaine sensitization: Behavioral and neurochemical aspects. *Neuroscience* **109,** 267–274.

Scheggi, S., Rauggi, R., Gambarana, C., Tagliamonte, A., and De Montis, M. G. (2004). Dopamine and cyclic AMP-regulated phosphoprotein-32 phosphorylation pattern in cocaine and morphine-sensitized rats. *J. Neurochem.* **90,** 792–799.

Schoepfer, R., Monyer, H., Sommer, B., Wisden, W., Sprengel, R., Kuner, T., Lomeli, H., Herb, A., Kohler, M., Burnashev, N., *et al.* (1994). Molecular biology of glutamate receptors. *Prog. Neurobiol.* **42,** 353–357.

Schumann, J., and Yaka, R. (2009). Prolonged withdrawal from repeated noncontingent cocaine exposure increases NMDA receptor expression and ERK activity in the nucleus accumbens. *J. Neurosci.* **29,** 6955–6963.

Sesack, S. R., and Grace, A. A. (2009). Cortico-basal ganglia reward network: Microcircuitry. *Neuropsychopharmacology*.

Shaham, Y., Shalev, U., Lu, L., De Wit, H., and Stewart, J. (2003). The reinstatement model of drug relapse: History, methodology and major findings. *Psychopharmacology (Berl)* **168,** 3–20.

Simoes, P. F., Silva, A. P., Pereira, F. C., Marques, E., Grade, S., Milhazes, N., Borges, F., Ribeiro, C. F., and Macedo, T. R. (2007). Methamphetamine induces alterations on hippocampal NMDA and AMPA receptor subunit levels and impairs spatial working memory. *Neuroscience* **150,** 433–441.

Smith, Y., and Villalba, R. (2008). Striatal and extrastriatal dopamine in the basal ganglia: An overview of its anatomical organization in normal and Parkinsonian brains. *Mov. Disord.* **23** (Suppl. 3), S534–S547.

Standaert, D. G., Landwehrmeyer, G. B., Kerner, J. A., Penney, J. B. Jr., and Young, A. B. (1996). Expression of NMDAR2D glutamate receptor subunit mRNA in neurochemically identified interneurons in the rat neostriatum, neocortex and hippocampus. *Brain Res. Mol. Brain Res.* **42,** 89–102.

Stipanovich, A., Valjent, E., Matamales, M., Nishi, A., Ahn, J. H., Maroteaux, M., Bertran-Gonzalez, J., Brami-Cherrier, K., Enslen, H., Corbille, A. G., Filhol, O., Nairn, A. C., *et al.* (2008). A phosphatase cascade by which rewarding stimuli control nucleosomal response. *Nature* **453,** 879–884.

Sun, X., Milovanovic, M., Zhao, Y., and Wolf, M. E. (2008). Acute and chronic dopamine receptor stimulation modulates AMPA receptor trafficking in nucleus accumbens neurons cocultured with prefrontal cortex neurons. *J. Neurosci.* **28,** 4216–4230.

Suzuki, T., and Okumura-Noji, K. (1995). NMDA receptor subunits epsilon 1 (NR2A) and epsilon 2 (NR2B) are substrates for Fyn in the postsynaptic density fraction isolated from the rat brain. *Biochem. Biophys. Res. Commun.* **216,** 582–588.

Svenningsson, P., Nishi, A., Fisone, G., Girault, J. A., Nairn, A. C., and Greengard, P. (2004). DARPP-32: An integrator of neurotransmission. *Annu. Rev. Pharmacol. Toxicol.* **44,** 269–296.

Thomas, G. M., and Huganir, R. L. (2004). MAPK cascade signalling and synaptic plasticity. *Nat. Rev. Neurosci.* **5,** 173–183.

Tzschentke, T. M. (2002). Glutamatergic mechanisms in different disease states: Overview and therapeutical implications—An introduction. *Amino Acids* **23,** 147–152.

Valjent, E., Corbille, A. G., Bertran-Gonzalez, J., Herve, D., and Girault, J. A. (2006). Inhibition of ERK pathway or protein synthesis during reexposure to drugs of abuse erases previously learned place preference. *Proc. Natl. Acad. Sci. USA* **103,** 2932–2937.

Valjent, E., Corvol, J. C., Pages, C., Besson, M. J., Maldonado, R., and Caboche, J. (2000). Involvement of the extracellular signal-regulated kinase cascade for cocaine-rewarding properties. *J. Neurosci.* **20,** 8701–8709.

Valjent, E., Pascoli, V., Svenningsson, P., Paul, S., Enslen, H., Corvol, J. C., Stipanovich, A., Caboche, J., Lombroso, P. J., Nairn, A. C., Greengard, P., Herve, D., *et al.* (2005). Regulation of a protein phosphatase cascade allows convergent dopamine and glutamate signals to activate ERK in the striatum. *Proc. Natl. Acad. Sci. USA* **102,** 491–496.

Voorn, P., Vanderschuren, L. J., Groenewegen, H. J., Robbins, T. W., and Pennartz, C. M. (2004). Putting a spin on the dorsal-ventral divide of the striatum. *Trends Neurosci.* **27,** 468–474.

Wang, J. Q., Liu, X., Zhang, G., Parelkar, N. K., Arora, A., Haines, M., Fibuch, E. E., and Mao, L. (2006). Phosphorylation of glutamate receptors: A potential mechanism for the regulation of receptor function and psychostimulant action. *J. Neurosci. Res.* **84,** 1621–1629.

Watanabe, M., Inoue, Y., Sakimura, K., and Mishina, M. (1993). Distinct distributions of five N-methyl-D-aspartate receptor channel subunit mRNAs in the forebrain. *J. Comp. Neurol.* **338,** 377–390.

Watanabe, M., Mishina, M., and Inoue, Y. (1994). Distinct spatiotemporal expressions of five NMDA receptor channel subunit mRNAs in the cerebellum. *J. Comp. Neurol.* **343,** 513–519.

Zheng, F., Gingrich, M. B., Traynelis, S. F., and Conn, P. J. (1998). Tyrosine kinase potentiates NMDA receptor currents by reducing tonic zinc inhibition. *Nat. Neurosci.* **1,** 185–191.

DECIPHERING RETT SYNDROME WITH MOUSE GENETICS, EPIGENOMICS, AND HUMAN NEURONS

Jifang Tao, Hao Wu, and Yi Eve Sun

Department of Molecular & Medical Pharmacology and Psychiatry & Behavioral Sciences, University of California, Los Angeles, USA

Mutations of *MECP2 (methyl-CpG binding protein 2)* cause Rett syndrome (RTT). Mouse genetics studies have demonstrated that the lack of functional MeCP2 in the central nervous system leads to RTT-like symptoms, which could be reversed upon MeCP2 restoration. MeCP2 recognizes methylated CpG dinucleotides and may interact with other chromatin remodeling proteins. Although traditionally thought to be a transcription repressor, MeCP2 may also be involved in transcription activation. With the development of new technologies, deciphering the role of MeCP2 on a genome-wide scale is important for understanding of the RTT disease mechanisms.

I. Introduction

Rett syndrome (RTT; MIM312750) is a major autism-spectrum neurological disorder affecting girls with an incidence of 1 in 10,000 female births. The girls usually seem to develop normally in the first 8–16 months before they enter a stage of stagnation. The clinical symptoms of RTT include deceleration of head growth, loss of speech and purposeful hand movements, irregular breathing patterns, seizures, and autistic features (Hagberg *et al.*, 1983; Nomura, 2005;

INTERNATIONAL REVIEW OF NEUROBIOLOGY, VOL. 89
DOI: 10.1016/S0074-7742(09)89007-7

Rett, 1966). Most male carriers of RTT mutations die perinatally, but there are RTT male cases reported with profound mental retardation and psychiatric disorders (Budden *et al.*, 2005; Dayer *et al.*, 2007; Jan *et al.*, 1999; Maiwald *et al.*, 2002; Masuyama *et al.*, 2005). In more than 95% classic RTT patients, *de novo* mutations of the X-chromosome-linked gene *MECP2 (methyl-CpG binding protein 2)* are found (Amir *et al.*, 1999; Wan *et al.*, 1999). In a small proportion of patients with RTT features with early-onset epilepsy and infantile spasms, mutations of the X-chromosome-linked gene *cyclin-dependent kinase like 5* (*CDKL5*) were identified (Grosso *et al.*, 2007). The *MECP2* mutations include missense, nonsense, frame-shift, and truncation mutations, which encompass whole exons (Christodoulou *et al.*, 2003) (Fig. 1). Autopsy fails to detect any obvious neurodegeneration or neuronal migration problems (Jellinger *et al.*, 1988), although the RTT patients' brains show reduced brain weight and volume, as well as sparse and short dendritic spines (Armstrong, 2005; Belichenko *et al.*, 1994).

II. RTT Mouse Models

To gain insights into the RTT etiology, several mouse models have been generated. Using the Cre-LoxP technology, *Mecp2*$^{-/y}$ male mice lacking exon 3 or both exons 3 and 4, as well as mice in which *Mecp2* is specifically deleted in the brain (Nestin-Cre), show severe neurological symptoms such as hypoactivity, body trembling, irregular breathing, and subsequent death around 8–10 weeks of age.

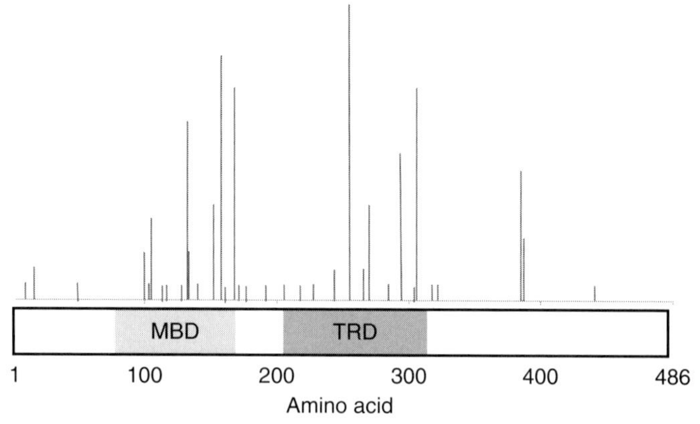

FIG. 1. Different RTT patient mutations frequency and their relative positions to the full length MeCP2 protein. MBD is the methyl-CpG binding domain and TRD is the transcription repression domain.

Mecp2$^{-/+}$ female mice initially show no symptoms until older than 3 months, when they start to exhibit reduced activity and ataxic gait. Deletion of *Mecp2* in postnatal CNS neurons (CamK-Cre93) leads to a similar neuronal phenotype, although at a later onset of around 3 months (Chen *et al.*, 2001; Guy *et al.*, 2001). These results suggest that MeCP2 plays essential roles in maintaining normal neuronal functions. Another mouse model with MeCP2 truncated at amino acid 308 (*Mecp2*$^{308/y}$), which mimics RTT patients with C-terminal deletions of the protein, exhibits progressive neurological phenotypes at a later onset age (Shahbazian *et al.*, 2002a). Also, twofold overexpression of MeCP2 under its endogenous promoter (*Mecp2*^{Tg}) or overexpression of MeCP2 in mature neurons with *tau* promoter results in progressive neurological abnormalities, suggesting that MeCP2 protein levels must be tightly regulated (Collins *et al.*, 2004; Luikenhuis *et al.*, 2004). Interestingly, restoration of MeCP2 expression in the symptomatic *Mecp2* null mice can effectively reverse some of the major phenotypes, suggesting that RTT is not strictly a neurodevelopmental disorder, but MeCP2 is required to stabilize and maintain the mature neuronal state (Guy *et al.*, 2007). This result also indicates that the molecular preconditions for normal MeCP2 activity are preserved in its absence. Recently, it is reported that systematic treatment of *Mecp2* mutant mice with an active peptide fragment of insulin-like growth factor 1 (IGF-1) could partially alleviate RTT-like symptoms (Tropea *et al.*, 2009).

III. MeCP2 Expressions and Functions

MeCP2 is ubiquitously expressed, with highest levels in the brain, especially in postmitotic neurons (Jung *et al.*, 2003). MeCP2 has two isoforms due to alternative splicing, which only differ in the N-terminus. The longer MeCP2 e1 (α) form is most abundant in the brain (Kriaucionis and Bird, 2004; Mnatzakanian *et al.*, 2004). In addition, MeCP2 has a long 3′UTR and its translation is regulated by a microRNA miR-132 (Klein *et al.*, 2007). MeCP2 protein levels are low in embryos and increase during neuronal maturation (Shahbazian *et al.*, 2002b).

MeCP2 is a member of the methyl-CpG binding protein family. It contains the methyl-CpG binding domain (MBD), the transcription repression domain (TRD) and the N- and C-terminal domains with unknown functions (Lewis *et al.*, 1992; Nan *et al.*, 1997). *In vitro*, MeCP2 recognizes methylated CpG dinucleotides with preference of adjacent A/T sequences through the MBD and inhibits gene transcription by recruiting corepressors including mSin3A and histone deacetylases (HDAC1 and HDAC2) via its TRD domain, potentially bridging DNA methylation and chromatin remodeling (Nan *et al.*, 1998). MeCP2 may also interact with ATRX, a SWI2/SNF2 DNA helicase/ATPase (Nan *et al.*, 2007).

In addition to its methylated DNA binding properties, MeCP2 might be involved in regulating RNA splicing as it interacts with RNA binding protein Y-box binding protein 1 (YB1), and aberrant RNA splicing patterns are observed in *Mecp2* mutant mouse brains (Young *et al.*, 2005). Also, its involvement in chromatin compaction and nucleosome clustering *in vitro* has been reported (Georgel *et al.*, 2003; Nikitina *et al.*, 2007). In terms of neurophysiological outcomes, MeCP2 is implicated in neuronal maturation (Kishi and Macklis, 2004) and synaptogenesis (Fukuda *et al.*, 2005). Lack of MeCP2 also shifts the balance of excitatory–inhibitory neurotransmission (Chao *et al.*, 2007; Dani *et al.*, 2005; Medrihan *et al.*, 2008; Nelson *et al.*, 2006). A recent study suggests that astrocytes in the RTT mice brains have a cell non-autonomous effect on neuronal dendritic morphology (Ballas *et al.*, 2009). However, the mechanistic link between MeCP2 DNA binding and neurophysiological functions is not yet clear.

IV. MeCP2 and Gene Expression

Given the hypothesized role of MeCP2 as a transcription regulator, it is critical to pursue the transcriptional deregulated target genes due to the absence of MeCP2. An initial attempt of transcription profiling comparing wild-type versus *Mecp2* null mice brain tissues yields only subtle gene expression changes (Tudor *et al.*, 2002).

Using a candidate gene approach, our group and others successfully identified *Bdnf* as the first mammalian neuronal target gene of MeCP2 (Chen *et al.*, 2003; Martinowich *et al.*, 2003). BDNF is a critical neurotrophic factor that is implicated in neuronal survival and adult neural plasticity (West *et al.*, 2001). The increased synthesis of BDNF in neurons after depolarization by potassium chloride correlates with dissociation of MeCP2–HDAC–mSin3A complexes from one of its promoters (Chen *et al.*, 2003; Martinowich *et al.*, 2003). Subsequent *in vivo* studies surprisingly show that BDNF levels in *Mecp2*$^{-/y}$ mice brains have decreased rather than increased. Furthermore, forebrain-specific *Bdnf* depletion in *Mecp2*$^{-/y}$ mice results in earlier onset of symptoms and reduced lifespan while overexpression of BDNF in *Mecp2* null mice somewhat alleviates the symptoms and extends the lifespan, again suggesting that BDNF misregulation is critically involved in RTT etiology (Chang *et al.*, 2006). Other genes identified by candidate approaches are involved in hormonal signaling, neuronal transcription, etc. as reviewed by Chahrour and Zoghbi (2007).

Recently, a large number of genes are identified by high-resolution exon microarrays (30–40 probes per gene) with moderate gene expression changes in the hypothalamus and cerebellum of mice that are either lack of or overexpressing MeCP2. Many of the altered genes are upregulated in MeCP2 overexpressed mice and downregulated in *Mecp2* mutant mice, suggesting either directly or indirectly, MeCP2 may function not only as a repressor but also as a transcription activator (Ben-Shachar *et al.*, 2009; Chahrour *et al.*, 2008). These surprising findings suggest that compared with previous microarrays designed mostly with probes near 3′ transcripts, the whole transcript exon array seems to be more sensitive to detect changes caused by MeCP2 protein level difference.

V. Posttranslational Modification of MeCP2

To add to the complexity of MeCP2 and neuronal-activity-dependent gene transcription, MeCP2 is found to undergo dynamic phosphorylation events in response to neuronal activity. One of the sites, serine 421 (S421), is induced by potassium chloride depolarization and implicated in regulating *Bdnf* transcription, dendritic growth, and spine maturation (Zhou *et al.*, 2006). Another site, serine 80 (S80), is negatively regulated by neuronal activity and may be important for MeCP2 chromatin association (Tao *et al.*, 2009). Mutations of either site phosphorylation *in vivo* by serine to alanine knock-in technique lead to locomotor abnormalities that are different from *Mecp2* null mice as summarized in Table I, suggesting that posttranslational modification of MeCP2 plays distinctive functional roles (Tao *et al.*, 2009).

VI. Epigenomics

Epigenetics refers to heritable traits caused by mechanisms other than changes in the DNA sequence. The molecular basis of epigenetics is complex as it includes DNA methylation, chromatin modifications, and non-coding RNA transcripts. The total epigenetic information is termed “epigenome.” Different from the genome, which is relatively stable during a eukaryotic organism’s life span, epigenome is constantly changing during differentiation, development, and aging. Therefore, there are many epigenomes versus one single genome. MeCP2 is a multifuncitonal chromatin-bound protein recognizing DNA methylation with potential interaction with chromatin remodeling complexes; it shall be scrutinized on a new genome-wide scale in the context of both DNA methylation and histone modifications.

Recently, high-throughput approaches have been developed to analyze DNA methylation patterns across the whole genome. Methylation-sensitive restriction enzyme or affinity purification methods (with either an antibody recognizing methylated cytosine or MBD domain of methyl-CpG binding protein) can be

TABLE I

COMPARISON OF *Mecp2* MUTANT MICE

	Mouse strain	Life span	Body and brain weights	Symptoms onset time	Phenotypes
Mecp2$^{-/y}$ by Jaenisch group (Chen *et al.*, 2001)	C57BL/6J, 129/Sv and Balb/c	Death around 10 weeks of age	Reduced brain weight and smaller neuronal cell sizes. Gain of body weight at early symptomatic stage and loss of weight at late stage.	5 weeks of age	Body trembling, hypoactive, hard respiration, ataxia gait.
Mecp2$^{-/y}$ by Bird group (Guy *et al.*, 2001)	C57BL/6J	Death around postnatal 54 days	Varying body weight	3–8 weeks of age	Stiff and uncoordinated gait, reduced spontaneous movement, hindlimb clasping, irregular breathing.
Mecp2lox/y;Nestin-Cre (Chen *et al.*, 2001)	C57BL/6J, 129/Sv and Balb/c	Death around 10 weeks of age	Reduced brain weight and smaller neuronal cell sizes. Gain of body weight at early symptomatic stage and loss of weight at late stage.	5 weeks of age	Body trembling, hypoactive, hard respiration, ataxia gait.
Mecp2lox/y;CamK –Cre (Chen *et al.*, 2001)	C57BL/6J, 129/Sv and Balb/c	>7.5 months	Reduced brain weight and smaller neuronal cell sizes. Gain of body weight at early symptomatic stage.	3 months	Ataxic gait and reduced nocturnal activity.
Mecp2 308/y (Shahbazian *et al.*, 2002a)	C57BL/6J, 129/Sv	Most survived to 1 year of age	Normal brain and body weight	Subtle tremor at 6 weeks of age; apparent at 4 months of age	Tremors, motor impairments, hypoactivity, increased anxiety-related behavior, seizures, kyphosis and stereotypic forelimb motions.
Mecp2 S80A/y (Tao *et al.*, 2009)	C57BL/6J, 129/Sv	Normal life span	Normal brain weight and slightly heavier body weight	2 to 3 months of age	Reduced locomotor activity.
Mecp2 S421AS424A/y (Tao *et al.*, 2009)	C57BL/6J, 129/Sv	Normal life span	Slightly heavier brain weight and normal body weight	2 to 3 months of age	Increased locomotor activity.

The table is adapted from supporting table 1 of Tao *et al.* (2009).

used to enrich the methylated DNA (Cross *et al.*, 1994; Frommer *et al.*, 1992; Irizarry *et al.*, 2008; Khulan *et al.*, 2006; Lippman *et al.*, 2005; Zhang *et al.*, 2006). The enriched samples are then either hybridized with microarray or directly subject to large-scale sequencing (Cokus *et al.*, 2008; Weber *et al.*, 2005; Zilberman *et al.*, 2007). Recent new findings from these genomic studies suggest that DNA methylation not only occurs at gene promoters but also exists in gene bodies of actively transcribed genes (Hellman and Chess, 2007; Suzuki and Bird, 2008). Thus, the function of DNA methylation may not be limited to transcription repression. What is more, although still controversial, several reports have suggested that DNA methylation may not be as stable as a permanent mark as it may be dynamically regulated (Jin *et al.*, 2008; Kangaspeska *et al.*, 2008; Ma *et al.*, 2009; Metivier *et al.*, 2008; Tahiliani *et al.*, 2009). Therefore, it is conceivable to speculate that the localization of MeCP2, one of the DNA methylation pattern readers, might also be widespread and dynamic. To survey MeCP2 genome-wide localization, the combination of chromatin immunoprecipitation (ChIP) and DNA microarrays (ChIP-chip) or massively parallel sequencing (ChIP-seq) shall be carried out with specific anti-MeCP2 antibodies and in a homogenous or highly enriched neuronal population (Fig. 2).

Besides DNA methylation, numerous genome-wide studies have provided data on the distribution of chromatin modifications in various cell types. Histone acetylations and histone H3 lysine4 trimethylation (H3K4me3) correlate positively with transcription levels, while H3K4me3 might also mark poised promoters (Bernstein *et al.*, 2005, 2006; Guenther *et al.*, 2007; Kim *et al.*, 2005; Roh *et al.*, 2005, 2006). H3K9me3, H3K27me3, and H4K20me3 have been implicated in transcription repression or heterochromatic domains (Boyer *et al.*, 2006; Martens *et al.*, 2005; Squazzo *et al.*, 2006). The knowledge about histone modifications and their functions (Kouzarides, 2007; Schones and Zhao, 2008) will be critical for studying the properties of MeCP2 binding regions.

VII. Human Neuron-Based RTT Disease Model

From previous RTT mouse models, it is noteworthy that human are more sensitive to MeCP2 dysfunction. Girls that are mosaic with mutant *MECP2* due to X-chromosome inactivation exhibit classic RTT phenotypes, while heterozygous female *Mecp2*$^{-/+}$ mice remain largely normal until ages more than 3 months. *MECP2* mutations in boys usually result in early death or profound mental retardations. *Mecp2*$^{-/y}$ mice show RTT-like phenotypes rather than perinatal death (Chen *et al.*, 2001; Guy *et al.*, 2001). Therefore, it is important to build a cellular RTT disease model with human neurons.

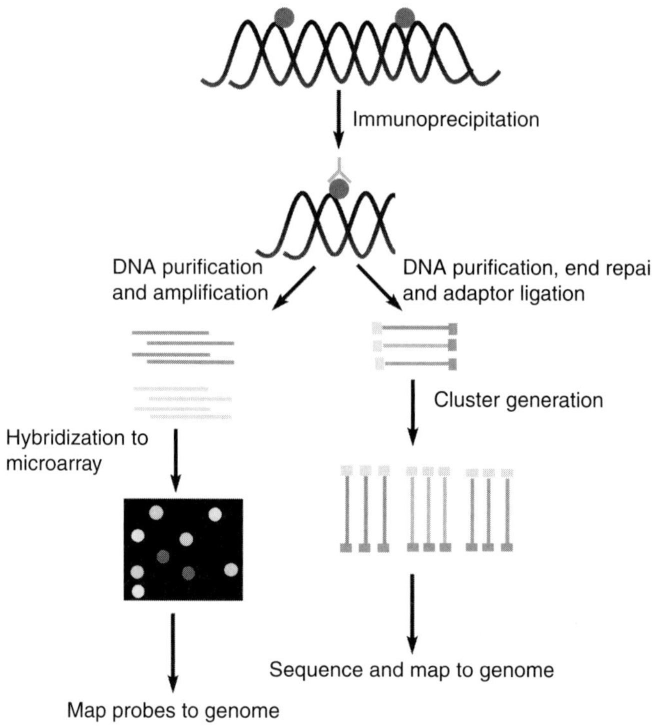

FIG. 2. Chromatin immunoprecipitation combined with DNA microarrays (ChIP-chip) or high throughput sequencing (ChIP-seq). The protein of interest (MeCP2, red ball) that is bound to chromatin is immunoprecipitated together with cross-linked DNA by a specific antibody (blue). The immunoprecipitated DNA is reverse cross-linked, purified and gone through amplification to obtain sufficient DNA and color labeled for DNA microarrays. Alternatively, for sequencing (Solexa shown here), the ChIP DNA ends are repaired and ligated to a pair of adaptors, followed by amplification PCR. The DNAs are immobilized on the surface of a flow cell that contains oligonucleotides that recognize the adaptor sequences. Clusters of DNA are sequenced during synthesis. The sequence reads are mapped to reference genome.

Human embryonic stem cells (hESCs) derived from human blastocysts are capable of self-renewal and proliferation, and have the potential to be differentiated into virtually all the cell types in the human body (Murry and Keller, 2008; Thomson *et al.*, 1998). Homologous recombination-mediated gene targeting (Xue *et al.*, 2009; Zou *et al.*, 2009; Zwaka and Thomson, 2003) or lentiviral-mediated manipulations can be used to generate hESCs with MeCP2 deficiency that mimics RTT patient mutations. hESCs can be differentiated into highly enriched human neurons that are able to form functional synapses (Wu *et al.*, 2007). The human neurons carrying RTT patients' mutations may be useful for investigating the disease mechanisms that lead to RTT and for identifying of small molecules for counteracting the RTT mutation effects.

Recently, the rapid evolving technology of induced pluripotent stem-cell-like (iPS) cells has offered unprecedented opportunities for disease modeling, drug screening, toxicology, and regenerative medicine (Dimos *et al.*, 2008; Park *et al.*, 2008; Takahashi and Yamanaka, 2006). Four transcription factors, Oct3/4 (also known as Pou5f1), Sox2, Klf4, and c-Myc, in modified combination or with facilitation of small molecules, e.g., valporic acid, have been introduced into various terminally differentiated cells by retroviruses, adenoviruses, lentiviruses, self-cleavage peptides, episomal DNA, and protein transductions and have successfully reprogrammed the lineage committed cells into stem-cell-like state (Huangfu *et al.*, 2008; Stadtfeld *et al.*, 2008; Woltjen *et al.*, 2009; Yu *et al.*, 2009; Zhou *et al.*, 2009). It is conceivable that different RTT-patient-specific iPS cells may be generated for comparison of cellular phenotypes and may lead to personalized drug screening (Hotta *et al.*, 2009).

VIII. Summary

In summary, RTT is an autism-spectrum disorder caused by mutations of a single gene. The mouse genetics studies have provided important insights about the physiological nature of RTT. The studies on MeCP2 protein itself and its candidate target genes have enriched our understandings of the MeCP2 functioning context. The unresolved central question, however, is how MeCP2 functions on a genome-wide scale. This will need comprehensive analysis integrating DNA methylation, chromatin modification, and transcription profiling. With human neurons derived from hESC or iPS cells, RTT cellular mechanism and drug discovery studies shall shed lights into our understanding of neurological disorders even beyond RTT.

References

Amir, R. E., Van den Veyver, I. B., Wan, M., Tran, C. Q., Francke, U., and Zoghbi, H. Y. (Eds.) (1999). Rett syndrome is caused by mutations in X-linked MECP2, encoding methyl-CpG-binding protein 2. *Nat. Genet.* **23,** 185–188.

Armstrong, D. D. (2005). Neuropathology of Rett syndrome. *J. Child Neurol.* **20,** 747–753.

Ballas, N., Lioy, D. T., Grunseich, C., and Mandel, G. (2009). Non-cell autonomous influence of MeCP2-deficient glia on neuronal dendritic morphology. *Nat. Neurosci.* **12,** 311–317.

Belichenko, P. V., Oldfors, A., Hagberg, B., and Dahlstrom, A. (1994). Rett syndrome: 3-D confocal microscopy of cortical pyramidal dendrites and afferents. *Neuroreport* **5,** 1509–1513.

Ben-Shachar, S., Chahrour, M., Thaller, C., Shaw, C. A., and Zoghbi, H. Y. (2009). Mouse models of MeCP2 disorders share gene expression changes in the cerebellum and hypothalamus. *Hum. Mol. Genet.* **18,** 2431–2442.

Bernstein, B. E., Kamal, M., Lindblad-Toh, K., Bekiranov, S., Bailey, D. K., Huebert, D. J., McMahon, S., Karlsson, E. K., Kulbokas, E. J. 3rd., Gingeras, T. R., Schreiber, S. L., and Lander, E. S. (2005). Genomic maps and comparative analysis of histone modifications in human and mouse. *Cell* **120,** 169–181.

Bernstein, B. E., Mikkelsen, T. S., Xie, X., Kamal, M., Huebert, D. J., Cuff, J., Fry, B., Meissner, A., Wernig, M., Plath, K., Jaenisch, R., Wagschal, A., *et al.* (2006). A bivalent chromatin structure marks key developmental genes in embryonic stem cells. *Cell* **125,** 315–326.

Boyer, L. A., Plath, K., Zeitlinger, J., Brambrink, T., Medeiros, L. A., Lee, T. I., Levine, S. S., Wernig, M., Tajonar, A., Ray, M. K., Bell, G. W., Otte, A. P., *et al.* (2006). Polycomb complexes repress developmental regulators in murine embryonic stem cells. *Nature* **441,** 349–353.

Budden, S. S., Dorsey, H. C., and Steiner, R. D. (2005). Clinical profile of a male with Rett syndrome. *Brain Dev.* **27**(Suppl. 1), S69–S71.

Chahrour, M., and Zoghbi, H. Y. (2007). The story of Rett syndrome: From clinic to neurobiology. *Neuron* **56,** 422–437.

Chahrour, M., Jung, S. Y., Shaw, C., Zhou, X., Wong, S. T., Qin, J., and Zoghbi, H. Y. (2008). MeCP2, a key contributor to neurological disease, activates and represses transcription. *Science* **320,** 1224–1229.

Chang, Q., Khare, G., Dani, V., Nelson, S., and Jaenisch, R. (2006). The disease progression of Mecp2 mutant mice is affected by the level of BDNF expression. *Neuron* **49,** 341–348.

Chao, H. T., Zoghbi, H. Y., and Rosenmund, C. (2007). MeCP2 controls excitatory synaptic strength by regulating glutamatergic synapse number. *Neuron* **56,** 58–65.

Chen, R. Z., Akbarian, S., Tudor, M., and Jaenisch, R. (2001). Deficiency of methyl-CpG binding protein-2 in CNS neurons results in a Rett-like phenotype in mice. *Nat. Genet.* **27,** 327–331.

Chen, W. G., Chang, Q., Lin, Y., Meissner, A., West, A. E., Griffith, E. C., Jaenisch, R., and Greenberg, M. E. (2003). Derepression of BDNF transcription involves calcium-dependent phosphorylation of MeCP2. *Science* **302,** 885–889.

Christodoulou, J., Grimm, A., Maher, T., and Bennetts, B. (2003). RettBASE: The IRSA MECP2 variation database—A new mutation database in evolution. *Hum. Mutat* **21,** 466–472.

Cokus, S. J., Feng, S., Zhang, X., Chen, Z., Merriman, B., Haudenschild, C. D., Pradhan, S., Nelson, S. F., Pellegrini, M., and Jacobsen, S. E. (2008). Shotgun bisulphite sequencing of the Arabidopsis genome reveals DNA methylation patterning. *Nature* **452,** 215–219.

Collins, A. L., Levenson, J. M., Vilaythong, A. P., Richman, R., Armstrong, D. L., Noebels, J. L., David Sweatt, J., and Zoghbi, H. Y. (2004). Mild overexpression of MeCP2 causes a progressive neurological disorder in mice. *Hum. Mol. Genet.* **13,** 2679–2689.

Cross, S. H., Charlton, J. A., Nan, X., and Bird, A. P. (1994). Purification of CpG islands using a methylated DNA binding column. *Nat. Genet.* **6,** 236–244.

Dani, V. S., Chang, Q., Maffei, A., Turrigiano, G. G., Jaenisch, R., and Nelson, S. B. (2005). Reduced cortical activity due to a shift in the balance between excitation and inhibition in a mouse model of Rett syndrome. *Proc. Natl. Acad. Sci. USA* **102,** 12560–12565.

Dayer, A. G., Bottani, A., Bouchardy, I., Fluss, J., Antonarakis, S. E., Haenggeli, C. A., and Morris, M. A. (2007). MECP2 mutant allele in a boy with Rett syndrome and his unaffected heterozygous mother. *Brain Dev.* **29,** 47–50.

Dimos, J. T., Rodolfa, K. T., Niakan, K. K., Weisenthal, L. M., Mitsumoto, H., Chung, W., Croft, G. F., Saphier, G., Leibel, R., Goland, R., Wichterle, H., Henderson, C. E., *et al.* (2008).

Induced pluripotent stem cells generated from patients with ALS can be differentiated into motor neurons. *Science* **321,** 1218–1221.

Frommer, M., McDonald, L. E., Millar, D. S., Collis, C. M., Watt, F., Grigg, G. W., Molloy, P. L., and Paul, C. L. (1992). A genomic sequencing protocol that yields a positive display of 5-methylcytosine residues in individual DNA strands. *Proc. Natl. Acad. Sci. USA* **89,** 1827–1831.

Fukuda, T., Itoh, M., Ichikawa, T., Washiyama, K., and Goto, Y. (2005). Delayed maturation of neuronal architecture and synaptogenesis in cerebral cortex of Mecp2-deficient mice. *J. Neuropathol. Exp. Neurol.* **64,** 537–544.

Georgel, P. T., Horowitz-Scherer, R. A., Adkins, N., Woodcock, C. L., Wade, P. A., and Hansen, J. C. (2003). Chromatin compaction by human MeCP2. Assembly of novel secondary chromatin structures in the absence of DNA methylation. *J. Biol. Chem.* **278,** 32181–32188.

Grosso, S., Brogna, A., Bazzotti, S., Renieri, A., Morgese, G., and Balestri, P. (2007). Seizures and electroencephalographic findings in CDKL5 mutations: Case report and review. *Brain Dev.* **29,** 239–242.

Guenther, M. G., Levine, S. S., Boyer, L. A., Jaenisch, R., and Young, R. A. (2007). A chromatin landmark and transcription initiation at most promoters in human cells. *Cell* **130,** 77–88.

Guy, J., Hendrich, B., Holmes, M., Martin, J. E., and Bird, A. (2001). A mouse Mecp2-null mutation causes neurological symptoms that mimic Rett syndrome. *Nat. Genet.* **27,** 322–326.

Guy, J., Gan, J., Selfridge, J., Cobb, S., and Bird, A. (2007). Reversal of neurological defects in a mouse model of Rett syndrome. *Science* **315,** 1143–1147.

Hagberg, B., Aicardi, J., Dias, K., and Ramos, O. (1983). A progressive syndrome of autism, dementia, ataxia, and loss of purposeful hand use in girls: Rett's syndrome: Report of 35 cases. *Ann. Neurol.* **14,** 471–479.

Hellman, A., and Chess, A. (2007). Gene body-specific methylation on the active X chromosome. *Science* **315,** 1141–1143.

Hotta, A., Cheung, A. Y., Farra, N., Vijayaragavan, K., Seguin, C. A., Draper, J. S., Pasceri, P., Maksakova, I. A., Mager, D. L., Rossant, J., Bhatia, M., and Ellis, J. (2009). Isolation of human iPS cells using EOS lentiviral vectors to select for pluripotency. *Nat. Methods* **6,** 370–376.

Huangfu, D., Maehr, R., Guo, W., Eijkelenboom, A., Snitow, M., Chen, A. E., and Melton, D. A. (2008). Induction of pluripotent stem cells by defined factors is greatly improved by small-molecule compounds. *Nat. Biotechnol.* **26,** 795–797.

Irizarry, R. A., Ladd-Acosta, C., Carvalho, B., Wu, H., Brandenburg, S. A., Jeddeloh, J. A., Wen, B., and Feinberg, A. P. (2008). Comprehensive high-throughput arrays for relative methylation (CHARM). *Genome Res* **18,** 780–790.

Jan, M. M., Dooley, J. M., and Gordon, K. E. (1999). Male Rett syndrome variant: Application of diagnostic criteria. *Pediatr. Neurol.* **20,** 238–240.

Jellinger, K., Armstrong, D., Zoghbi, H. Y., and Percy, A. K. (1988). Neuropathology of Rett syndrome. *Acta Neuropathol.* **76,** 142–158.

Jin, S. G., Guo, C., and Pfeifer, G. P. (2008). GADD45A does not promote DNA demethylation. *PLoS Genet.* **e1000013,** 4.

Jung, B. P., Jugloff, D. G., Zhang, G., Logan, R., Brown, S., and Eubanks, J. H. (2003). The expression of methyl CpG binding factor MeCP2 correlates with cellular differentiation in the developing rat brain and in cultured cells. *J. Neurobiol.* **55,** 86–96.

Kangaspeska, S., Stride, B., Metivier, R., Polycarpou-Schwarz, M., Ibberson, D., Carmouche, R. P., Benes, V., Gannon, F., and Reid, G. (2008). Transient cyclical methylation of promoter DNA. *Nature* **452,** 112–115.

Khulan, B., Thompson, R. F., Ye, K., Fazzari, M. J., Suzuki, M., Stasiek, E., Figueroa, M. E., Glass, J. L., Chen, Q., Montagna, C., Hatchwell, E., and Selzer, R. R. (2006). Comparative isoschizomer profiling of cytosine methylation: The HELP assay. *Genome Res.* **16,** 1046–1055.

Kim, T. H., Barrera, L. O., Zheng, M., Qu, C., Singer, M. A., Richmond, T. A., Wu, Y., Green, R. D., and Ren, B. (2005). A high-resolution map of active promoters in the human genome. *Nature* **436,** 876–880.

Kishi, N., and Macklis, J. D. (2004). MECP2 is progressively expressed in post-migratory neurons and is involved in neuronal maturation rather than cell fate decisions. *Mol. Cell Neurosci.* **27,** 306–321.

Klein, M. E., Lioy, D. T., Ma, L., Impey, S., Mandel, G., and Goodman, R. H. (2007). Homeostatic regulation of MeCP2 expression by a CREB-induced microRNA. *Nat. Neurosci.* **10,** 1513–1514.

Kouzarides, T. (2007). Chromatin modifications and their function. *Cell* **128,** 693–705.

Kriaucionis, S., and Bird, A. (2004). The major form of MeCP2 has a novel N-terminus generated by alternative splicing. *Nucleic Acids Res.* **32,** 1818–1823.

Lewis, J. D., Meehan, R. R., Henzel, W. J., Maurer-Fogy, I., Jeppesen, P., Klein, F., and Bird, A. (1992). Purification, sequence, and cellular localization of a novel chromosomal protein that binds to methylated DNA. *Cell* **69,** 905–914.

Lippman, Z., Gendrel, A. V., Colot, V., and Martienssen, R. (2005). Profiling DNA methylation patterns using genomic tiling microarrays. *Nat. Methods* **2,** 219–224.

Luikenhuis, S., Giacometti, E., Beard, C. F., and Jaenisch, R. (2004). Expression of MeCP2 in postmitotic neurons rescues Rett syndrome in mice. *Proc. Natl. Acad. Sci. USA* **101,** 6033–6038.

Ma, D. K., Jang, M. H., Guo, J. U., Kitabatake, Y., Chang, M. L., Pow-Anpongkul, N., Flavell, R. A., Lu, B., Ming, G. L., and Song, H. (2009). Neuronal activity-induced Gadd45b promotes epigenetic DNA demethylation and adult neurogenesis. *Science* **323,** 1074–1077.

Maiwald, R., Bonte, A., Jung, H., Bitter, P., Storm, Z., Laccone, F., and Herkenrath, P. (2002). *De novo* MECP2 mutation in a 46,XX male patient with Rett syndrome. *Neurogenetics* **4,** 107–108.

Martens, J. H., O'Sullivan, R. J., Braunschweig, U., Opravil, S., Radolf, M., Steinlein, P., and Jenuwein, T. (2005). The profile of repeat-associated histone lysine methylation states in the mouse epigenome. *EMBO J.* **24,** 800–812.

Martinowich, K., Hattori, D., Wu, H., Fouse, S., He, F., Hu, Y., Fan, G., and Sun, Y. E. (2003). DNA methylation-related chromatin remodeling in activity-dependent BDNF gene regulation. *Science* **302,** 890–893.

Masuyama, T., Matsuo, M., Jing, J. J., Tabara, Y., Kitsuki, K., Yamagata, H., Kan, Y., Miki, T., Ishii, K., and Kondo, I. (2005). Classic Rett syndrome in a boy with R133C mutation of MECP2. *Brain Dev.* **27,** 439–442.

Medrihan, L., Tantalaki, E., Aramuni, G., Sargsyan, V., Dudanova, I., Missler, M., and Zhang, W. (2008). Early defects of GABAergic synapses in the brain stem of a MeCP2 mouse model of Rett syndrome. *J. Neurophysiol.* **99,** 112–121.

Metivier, R., Gallais, R., Tiffoche, C., Le Peron, C., Jurkowska, R. Z., Carmouche, R. P., Ibberson, D., Barath, P., Demay, F., Reid, G., Benes, V., and Jeltsch, A. (2008). Cyclical DNA methylation of a transcriptionally active promoter. *Nature* **452,** 45–50.

Mnatzakanian, G. N., Lohi, H., Munteanu, I., Alfred, S. E., Yamada, T., MacLeod, P. J., Jones, J. R., Scherer, S. W., Schanen, N. C., Friez, M. J., Vincent, J. B., and Minassian, B. A. (2004). A previously unidentified MECP2 open reading frame defines a new protein isoform relevant to Rett syndrome. *Nat. Genet.* **36,** 339–341.

Murry, C. E., and Keller, G. (2008). Differentiation of embryonic stem cells to clinically relevant populations: Lessons from embryonic development. *Cell* **132,** 661–680.

Nan, X., Campoy, F. J., and Bird, A. (1997). MeCP2 is a transcriptional repressor with abundant binding sites in genomic chromatin. *Cell* **88,** 471–481.

Nan, X., Ng, H. H., Johnson, C. A., Laherty, C. D., Turner, B. M., Eisenman, R. N., and Bird, A. (1998). Transcriptional repression by the methyl-CpG-binding protein MeCP2 involves a histone deacetylase complex. *Nature* **393,** 386–389.

Nan, X., Hou, J., Maclean, A., Nasir, J., Lafuente, M. J., Shu, X., Kriaucionis, S., and Bird, A. (2007). Interaction between chromatin proteins MECP2 and ATRX is disrupted by mutations that cause inherited mental retardation. *Proc. Natl. Acad. Sci. USA* **104,** 2709–2714.

Nelson, E. D., Kavalali, E. T., and Monteggia, L. M. (2006). MeCP2-dependent transcriptional repression regulates excitatory neurotransmission. *Curr. Biol.* **16,** 710–716.

Nikitina, T., Shi, X., Ghosh, R. P., Horowitz-Scherer, R. A., Hansen, J. C., and Woodcock, C. L. (2007). Multiple modes of interaction between the methylated DNA binding protein MeCP2 and chromatin. *Mol. Cell Biol.* **27,** 864–877.

Nomura, Y. (2005). Early behavior characteristics and sleep disturbance in Rett syndrome. *Brain Dev.* **27**(Suppl. 1), S35–S42.

Park, I. H., Arora, N., Huo, H., Maherali, N., Ahfeldt, T., Shimamura, A., Lensch, M. W., Cowan, C., Hochedlinger, K., and Daley, G. Q. (2008). Disease-specific induced pluripotent stem cells. *Cell* **134,** 877–886.

Rett, A. (1966). [On a unusual brain atrophy syndrome in hyperammonemia in childhood]. *Wien Med. Wochenschr* **116,** 723–726.

Roh, T. Y., Cuddapah, S., and Zhao, K. (2005). Active chromatin domains are defined by acetylation islands revealed by genome-wide mapping. *Genes Dev.* **19,** 542–552.

Roh, T. Y., Cuddapah, S., Cui, K., and Zhao, K. (2006). The genomic landscape of histone modifications in human T cells. *Proc. Natl. Acad. Sci. USA* **103,** 15782–15787.

Schones, D. E., and Zhao, K. (2008). Genome-wide approaches to studying chromatin modifications. *Nat. Rev. Genet.* **9,** 179–191.

Shahbazian, M., Young, J., Yuva-Paylor, L., Spencer, C., Antalffy, B., Noebels, J., Armstrong, D., Paylor, R., and Zoghbi, H. (2002a). Mice with truncated MeCP2 recapitulate many Rett syndrome features and display hyperacetylation of histone H3. *Neuron* **35,** 243–254.

Shahbazian, M. D., Antalffy, B., Armstrong, D. L., and Zoghbi, H. Y. (2002b). Insight into Rett syndrome: MeCP2 levels display tissue- and cell-specific differences and correlate with neuronal maturation. *Hum. Mol. Genet.* **11,** 115–124.

Squazzo, S. L., Antalffy, B., Armstrong, D. L., and Zoghbi, H. Y. (2006). Suz12 binds to silenced regions of the genome in a cell-type-specific manner. *Genome Res.* **16,** 890–900.

Stadtfeld, M., Nagaya, M., Utikal, J., Weir, G., and Hochedlinger, K. (2008). Induced pluripotent stem cells generated without viral integration. *Science* **322,** 945–949.

Suzuki, M. M., and Bird, A. (2008). DNA methylation landscapes: Provocative insights from epigenomics. *Nat. Rev. Genet.* **9,** 465–476.

Tahiliani, M., Koh, K. P., Shen, Y., Pastor, W. A., Bandukwala, H., Brudno, Y., Agarwal, S., Iyer, L. M., Liu, D. R., Aravind, L., and Rao, A. (2009). Conversion of 5-methylcytosine to 5-hydroxymethylcytosine in mammalian DNA by MLL partner TET1. *Science* **324,** 930–935.

Takahashi, K., and Yamanaka, S. (2006). Induction of pluripotent stem cells from mouse embryonic and adult fibroblast cultures by defined factors. *Cell* **126,** 663–676.

Tao, J., Hu, K., Chang, Q., Wu, H., Sherman, N. E., Martinowich, K., Klose, R. J., Schanen, C., Jaenisch, R., Wang, W., and Sun, Y. E. (2009). Phosphorylation of MeCP2 at Serine 80 regulates its chromatin association and neurological function. *Proc. Natl. Acad. Sci. USA* **106,** 4882–4887.

Thomson, J. A., Itskovitz-Eldor, J., Shapiro, S. S., Waknitz, M. A., Swiergiel, J. J., Marshall, V. S., and Jones, J. M. (1998). Embryonic stem cell lines derived from human blastocysts. *Science* **282,** 1145–1147.

Tropea, D., Giacometti, E., Wilson, N. R., Beard, C., McCurry, C., Fu, D. D., Flannery, R., Jaenisch, R., and Sur, M. (2009). Partial reversal of Rett syndrome-like symptoms in MeCP2 mutant mice. *Proc. Natl. Acad. Sci. USA* **106,** 2029–2034.

Tudor, M., Akbarian, S., Chen, R. Z., and Jaenisch, R. (2002). Transcriptional profiling of a mouse model for Rett syndrome reveals subtle transcriptional changes in the brain. *Proc. Natl. Acad. Sci. USA* **99,** 15536–15541.

Wan, M., Lee, S. S., Zhang, X., Houwink-Manville, I., Song, H. R., Amir, R. E., Budden, S., Naidu, S., Pereira, J. L., Lo, I. F., Zoghbi, H. Y., Schanen, N. C., *et al.* (1999). Rett syndrome and beyond: Recurrent spontaneous and familial MECP2 mutations at CpG hotspots. *Am. J. Hum. Genet.* **65,** 1520–1529.

Weber, M., Davies, J. J., Wittig, D., Oakeley, E. J., Haase, M., Lam, W. L., and Schubeler, D. (2005). Chromosome-wide and promoter-specific analyses identify sites of differential DNA methylation in normal and transformed human cells. *Nat. Genet.* **37,** 853–862.

West, A. E., Chen, W. G., Dalva, M. B., Dolmetsch, R. E., Kornhauser, J. M., Shaywitz, A. J., Takasu, M. A., Tao, X., and Greenberg, M. E. (2001). Calcium regulation of neuronal gene expression. *Proc. Natl. Acad. Sci. USA* **98,** 11024–11031.

Woltjen, K., Michael, I. P., Mohseni, P., Desai, R., Mileikovsky, M., Hamalainen, R., Cowling, R., Wang, W., Liu, P., Gertsenstein, M., Kaji, K., Sung, H. K., *et al.* (2009). piggyBac transposition reprograms fibroblasts to induced pluripotent stem cells. *Nature* **458,** 766–770.

Wu, H., Xu, J., Pang, Z. P., Ge, W., Kim, K. J., Blanchi, B., Chen, C., Sudhof, T. C., and Sun, Y. E. (2007). Integrative genomic and functional analyses reveal neuronal subtype differentiation bias in human embryonic stem cell lines. *Proc. Natl. Acad. Sci. USA* **104,** 13821–13826.

Xue, H., Wu, S., Papadeas, S. T., Spusta, S., Swistowska, A. M., Macarthur, C. C., Mattson, M. P., Maragakis, N. J., Capecchi, M. R., Rao, M. S., Zeng, X., and Liu, Y. (2009). A targeted neuroglial reporter line generated by homologous recombination in human embryonic stem cells. *Stem Cells* **27,** 1836–2846.

Young, J. I., Hong, E. P., Castle, J. C., Crespo-Barreto, J., Bowman, A. B., Rose, M. F., Kang, D., Richman, R., Johnson, J. M., Berget, S., and Zoghbi, H. Y. (2005). Regulation of RNA splicing by the methylation-dependent transcriptional repressor methyl-CpG binding protein 2. *Proc. Natl. Acad. Sci. USA* **102,** 17551–17558.

Yu, J., Hu, K., Smuga-Otto, K., Tian, S., Stewart, R., Slukvin, II., and Thomson, J. A. (2009). Human induced pluripotent stem cells free of vector and transgene sequences. *Science* **324,** 797–801.

Zhang, X., Yazaki, J., Sundaresan, A., Cokus, S., Chan, S. W., Chen, H., Henderson, I. R., Shinn, P., Pellegrini, M., Jacobsen, S. E., and Ecker, J. R. (2006). Genome-wide high-resolution mapping and functional analysis of DNA methylation in arabidopsis. *Cell* **126,** 1189–1201.

Zhou, Z., Hong, E. J., Cohen, S., Zhao, W. N., Ho, H. Y., Schmidt, L., Chen, W. G., Lin, Y., Savner, E., Griffith, E. C., Hu, L., Steen, J. A., *et al.* (2006). Brain-specific phosphorylation of MeCP2 regulates activity-dependent Bdnf transcription, dendritic growth, and spine maturation. *Neuron* **52,** 255–269.

Zhou, H., Wu, S., Joo, J. Y., Zhu, S., Han, D. W., Lin, T., Trauger, S., Bien, G., Yao, S., Zhu, Y., Siuzdak, G., Scholer, H. R., *et al.* (2009). Generation of induced pluripotent stem cells using recombinant proteins. *Cell Stem Cell* **4,** 381–384.

Zilberman, D., Gehring, M., Tran, R. K., Ballinger, T., and Henikoff, S. (2007). Genome-wide analysis of *Arabidopsis thaliana* DNA methylation uncovers an interdependence between methylation and transcription. *Nat. Genet.* **39,** 61–69.

Zou, J., Maeder, M. L., Mali, P., Pruett-Miller, S. M., Thibodeau-Beganny, S., Chou, B. K., Chen, G., Ye, Z., Park, I. H., Daley, G. Q., Porteus, M. H., Joung, J. K., *et al.* (2009). Gene targeting of a disease-related gene in human induced pluripotent stem and embryonic stem cells. *Cell Stem Cell* **5,** 97–110.

Zwaka, T. P., and Thomson, J. A. (2003). Homologous recombination in human embryonic stem cells. *Nat. Biotechnol.* **21,** 319–321.

INDEX

CONTENTS OF RECENT VOLUMES

Volume 39

Volume 40

Volume 41

Volume 42

Volume 43

Volume 44

Volume 45

Volume 46

Volume 47

Volume 48

Volume 49

Volume 50

Volume 51

Volume 52

Volume 53

Volume 54

Volume 55

Volume 56

Volume 57

Volume 58

Volume 59

Volume 60

Volume 61

Volume 62

Volume 63

Volume 64

Volume 65

Volume 66

Volume 67

Volume 68

Volume 72

Volume 73

Volume 74

Volume 75

Volume 76

Volume 77

Volume 78

Volume 79

Volume 79

Volume 81

Volume 82

Volume 83

Volume 84

Volume 85

Volume 86

Volume 87

Volume 88

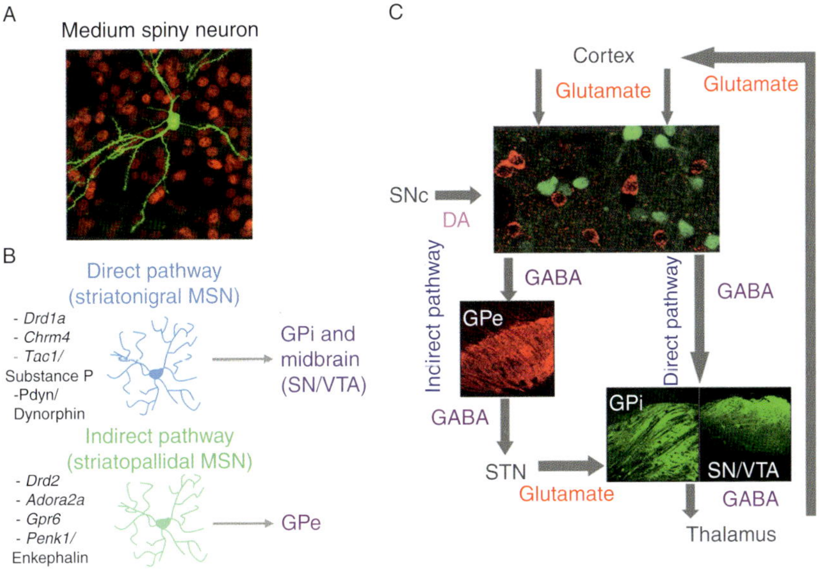

Mary Kay Lobo, Chapter 1, Fig. 1. (See Legend in Text.)

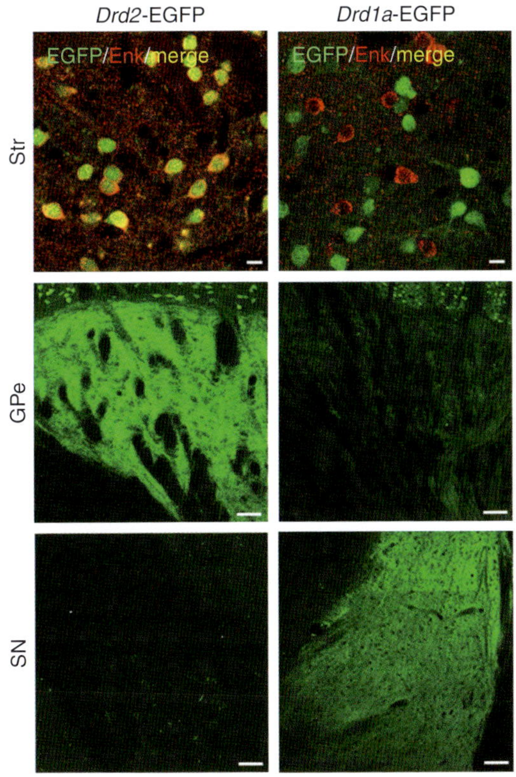

Mary Kay Lobo, Chapter 1, Fig. 3. (See Legend in Text.)

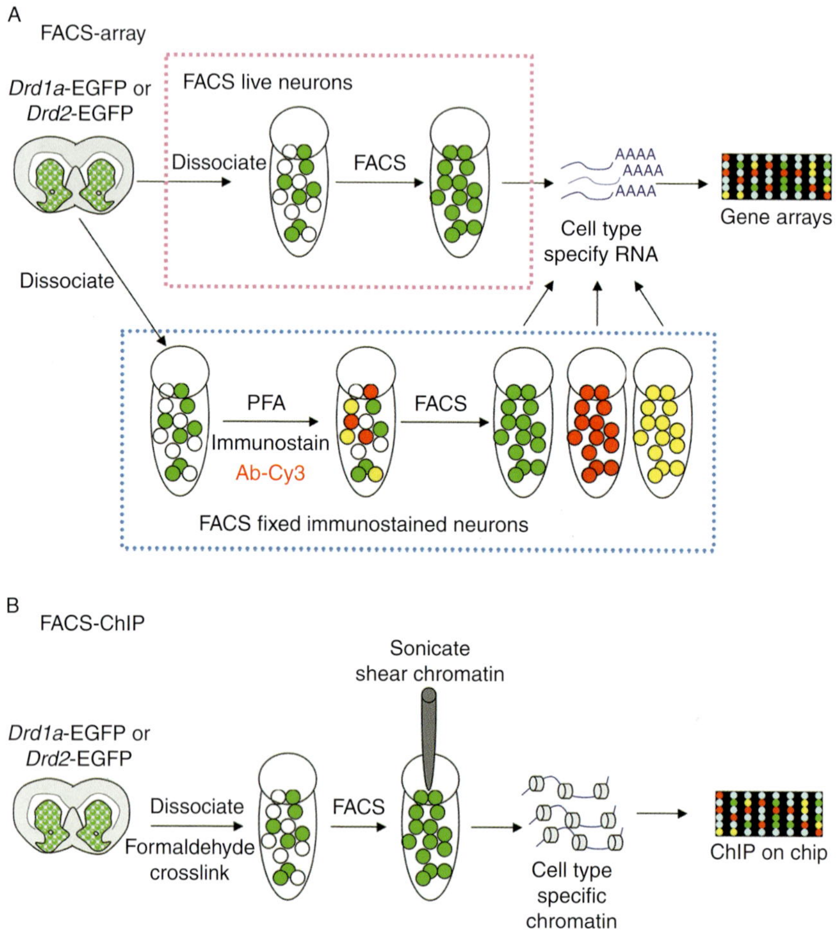

MARY KAY LOBO, CHAPTER 1, FIG. 4. (See Legend in Text.)